New Normal in Digital Enterprises

Subhra R. Mondal · Ceren Yegen ·
Subhankar Das
Editors

New Normal in Digital Enterprises

Strategies for Sustainable Development

palgrave
macmillan

Editors
Subhra R. Mondal
The Honors Programme
Southern Star Institute
of Management
Duy Tan University
Da Nang, Vietnam

Subhankar Das
The Honors Programme
Southern Star Institute
of Management
Duy Tan University
Da Nang, Vietnam

Ceren Yegen
Faculty of Communication
Mersin University
Mersin, Turkey

ISBN 978-981-19-8617-8 ISBN 978-981-19-8618-5 (eBook)
https://doi.org/10.1007/978-981-19-8618-5

This Palgrave Macmillan imprint is published by the registered company Springer Nature Singapore Pte Ltd.
The registered company address is: 152 Beach Road, #21-01/04 Gateway East, Singapore 189721, Singapore

Contents

Notes on Contributors

Asst. Prof. Dr. Gülay Asit was born on June 5, 1973, in Turkey. She graduated from Anadolu University, Turkey, with her B.A. degree in Business in 2012. She received M.A. degree in Public Relations from Girne American University, North Cyprus, in 2015. She then pursued her Ph.D. studies and received her Doctorate degree in Communication and Media from Girne American University in 2019. Since September 2019, she has been teaching as a full-time faculty member at the Faculty of Communication of Girne American University in North Cyprus. In February 2020, she was appointed as the Head of Digital Marketing and Social Media Department of the Faculty of Communication. Her research interests are New Media, Journalism, and Political Communication.

Mikail Batu got his Ph.D. degree from Ege University, Department of Public Relations in Turkey in 2012. After getting his Ph.D. degree, he did research on campaign management and social responsibility at Westminster Business School-London between 2014 and 2015. His main fields of study are corporate communication, social media, and campaign management. He has presented many papers at both international and national conferences on these subjects and all of his papers have been published as proceedings. Besides, he has articles published in national and international refereed journals. He has been working as an Associated Professor at Ege University, Faculty of Communication since 2018.

Ahmet Çetinkaya Ph.D., is an associate professor at Faculty of Communication at Marmara University in Istanbul, Turkey. His research interests include information management, decision support systems, information security, new media, and social media studies. He has published work in national and international indexed and refereed journals.

Subhankar Das (Academician, Researcher, Author, Writer, Blogger, Data Science trainer, and Social Media Marketing Consultant) received his Ph.D. in Social Media Advertisement and Media Planning. He has 12+ Years of Teaching and Research Experience. He is currently Professor and Researcher in the Honors Program at Duy Tan University, Da Nang, Vietnam. He has published 20+ Research Papers in various International Referred Journals (Scopus, Thomson Reuters, SSCI, SCIE).

Francesca Di Virgilio, Ph.D., is a Tenured Associate Professor of Organization Design and Human Resources Management in the Department of Economics at the University of Molise (Italy). National Habilitation as Full Professor of Organization Design and Human Resource Management from the Italian Ministry of University and Research. She is a member of the Association of Italian Organization Studies Academics (ASSIOA). She is a team member of the International academic scientist of the Spinner Innovation center (Research Excellence in Digital innovation. Lisboa—Portugal). She has been published in national and international academic journals, and she has contributed more chapters to various edited books. Her current research focuses on human resources management, organizational behavior, and knowledge management.

Olha Harmatiy Ph.D., is an Associate Professor for *Journalism and Mass Communication Department,* Lviv Polytechnic National University, Ukraine. The author has good publications in reputed journals and books.

Ali Murat Kirik is an Associate Professor at Marmara University Faculty of Communication, Department of Radio, Television and Cinema. In 2008, he obtained a Bachelor's Degree from Marmara University Faculty of Communication, Department of Radio, Television and Cinema. Meanwhile, he has worked in several media associations. In 2013, he was awarded a Ph.D. degree by the same department. His doctoral thesis is based on research he conducted on the use of Social Media by Turkish youth. His research interests include social media, digital communication, new media, and television broadcasting.

Özlem Kutlu graduated from Ege University, Faculty of Communication, Department of Public Relations and Publicity in 2005. She graduated from the Department of Business Administration in 2007 as her second undergraduate program. Kutlu received her master's degree from Ege University Social Sciences Institute in 2019 with the thesis title "Perceived Image of Political Party Leaders in the Turkish Grand National Assembly: A Research for State Universities in İzmir" took in. Kutlu, who won awards in many national competitions during her undergraduate education, works as an administrative staff at Ege University. Kutlu also works as a consultant in the field of Corporate Communications and Digital Marketing at Ajans Bee.

Jana Majerova Ph.D., works as an academic and research staff at AMBIS University, a.s. (Prague, Czech Republic) and as Associate Professor at the University of Zilina (Zilina, Slovak Republic). Her professional and research activities are focused on the issue of the Sustainable brand management and bankruptcy law. She has authored more than 170 publications: 4 scientific monographs; 3 university textbooks; 1 script; 36 publications in journals registered in WoS or SCOPUS, and more than 80 papers published in conference proceedings.

Subhra R. Mondal (Academician, Researcher, Author, and Writer) is a researcher in the Honors Program at Duy Tan University. She received her Ph.D. from the faculty of management, IBCS, SoA University. Her research interests are brand management, brand equity with innovative tools, smart tourism, destination marketing, social media marketing, AI & Web 5.0, and digital recruitment. She has over 12 years of experience in research and teaching. She is associated with different collaborative projects in European and Vietnamese organizations.

John Ben Prince is working as Assistant Professor at Symbiosis Institute of Business Management Bangalore. His corporate work experience spans ten years in industries such as Automobiles and Apparel. He has also spent about nine years in academics where he has been teaching subjects such as Strategic Management and Operations Management. He completed his Ph.D. at the Indian Institute of Management Lucknow. He has published in reputed journals and has attended international conferences.

Dr. Eliza Sharma is working with Symbiosis International University, Bengaluru, as Assistant Professor. She holds a Ph.D. in Management and Finance from Jaypee Institute of Information Technology, Noida. Her research work was on "Performance of Indian Commercial Banks and Its Relationship with Human Aspects in Banking". She has ten years of work experience in academics and research. She has published 35 research papers in national and international journals and conference proceedings. Her area of research is CSR, sustainability, ESG, SDG, and social issues.

Dr. Smriti Tandon is working as Associate Professor in the department of management studies at Graphic Era Deemed to be University, Dehradun. Her specialization is accounting and Finance. She is having 9 years of teaching experience.

Onur Tos graduated from Ege University, Faculty of Communication, Department of Public Relations and Publicity with the first place in 2015. Tos received his master's degree from Usak University Social Sciences Institute in 2018 with the thesis title "Perceived Image of Political Party Leaders in the Turkish Grand National Assembly: A Research for State Universities in İzmir" took in. Since 2019, Tos has been continuing his education at Ege University Institute of Social Sciences, Department of Public Relations and Publicity, Communication Research Doctorate Program. From 22 December 2020, he started to work as an Instructor in Tarsus University Vocational School Marketing and Advertising Department. He was appointed Head of Marketing and Advertising Department in February 2021.

Vasiliki Vrana is Professor at the Department of Business Administration, International Hellenic University, Greece, and serves as a tutor at the Hellenic Open University. She is widely published, with two books and more than 150 papers in academic journals and chapters in edited books and international conferences. Her research interests include Information and Communication Technologies in tourism and hospitality, education assessment, and Web 2.0 applications in tourism, health, and politics.

Ceren Yegen is an Associate Professor at Mersin University in Turkey. She graduated from Girne American University (GAU), Faculty of Communication, Department of Journalism and Broadcasting, and completed her master's degree and doctorate at Gazi University Social Sciences Institute, Journalism Department. She has many articles in various national and international journals. She also has chapters in new media and political communication themed books. Her research interests include nationalism, political communication, new media, and journalism.

List of Figures

List of Tables

PART I

Sustainable Digital Practice and COVID-19

CHAPTER 1

Exploration of Fake News and Their Impact on Sustainable Behavior in Controlling COVID-19

Gülay Asit

Introduction

The Sars-Cov2 virus, which emerged in Wuhan, China, spread all over the world in a short time, and as a result of the developments, the World Health Organization declared a pandemic. At of the end of January 2022, more than 375 million cases were confirmed and more than 5.5 million people have died. The number of vaccinations against the virus has approached to 10 billion (JHU, 2022). The pandemic has changed many aspects of our lives such as communication, working conditions, and socialization. The new communication technologies have become more important than ever at this stage. Especially internet and social networks have gained critical importance during the COVID-19 pandemic. The fact that life came to a standstill with the pandemic, along with long

G. Asit (✉)
Cyprus International University, Cyprus, Northern Cyprus
e-mail: gasit@ciu.edu.tr

S. R. Mondal et al. (eds.), *New Normal in Digital Enterprises*,
https://doi.org/10.1007/978-981-19-8618-5_1

periods of lockdown, required individuals to establish their connection with the outside world through these networks. While the severity of the pandemic and the speed of its spread all over the world caused people to panic, the scarcity of information about the virus increased this panic even more. The rapidly increasing number of deaths has been the scene of terrible images in some countries, which has further fueled anxiety and fear. Fortunately, the initiation of vaccination studies at some point slowed the course of the epidemic and was effective in calming the chaotic environment. However, in this process, a great deal of fake news about the disease and vaccines was circulated on the internet.

While the process was challenging enough for the whole world in every sense, the spread of false information and fake news made the situation even more difficult. The internet, especially social media sites, played a major role in the dissemination of this news. Unlike the traditional media, the lack of media professionalism and a supervisory mechanism has caused such news to circulate uncontrollably and rapidly on social media sites. As a result, the level of development of information technologies has ensured the rapid spread of falsehood and misinformation under the name of freedom of speech and has brought the world into an infodemic era (Eysenbach, 2020).

With the COVID-19 pandemic, the increase in the amount of fake news has reached incredible proportions. Social networks such as Facebook, Twitter, WhatsApp, and TikTok, which are widely used all over the world, played a major role in this increase. In the dissemination of fake news, the statements of politicians as well as ordinary people came to the fore. For example, in March 2020, the statement that the virus was a hoax of China was frequently voiced by politicians (Buchanan, 2020). In this process, the sharing of unverified content spread by clicking a single button has led to a worldwide epidemic of fake content, causing lynching and violence in some countries (Laskar & Reyaz, 2021).

Vaccination is of great importance to end the pandemic. The decrease in virus-related deaths and easier recovery from the disease depend on it. On the other hand, vaccine controversy is one of the barriers to vaccination and delays taking the pandemic under control. Because a lot of baseless news and content about vaccines and vaccination are circulating on the internet. Although many studies have been carried out on fake news about the COVID-19 virus, no research has yet been conducted on fake news about vaccines. This study aims to contribute to the field in this regard. In this context, firstly, the concept of infodemic and the role of social media are explained. The emergence of the concept, the types of fake news about COVID-19, and the reasons for the spread of fake news are mentioned. Next, fact-checking of the data on COVID-19 vaccines made by the fact-checking sites in Turkey, namely Doğruluk Payı, Doğrula, Malumatfuruş, and Teyit, were analyzed using the content analysis technique.

Literature Review

Infodemic

The spread of fake news about the COVID-19 pandemic and the efforts to prevent it have revealed the concept of infodemic. The concept refers to the excessive amount of information pollution circulated about the pandemic in social media (Öztürk, 2021). However, according to Carlos Navarro, head of Public Health Emergencies at UNICEF, the children's agency, while a lot of incorrect information is spreading through social media, a lot is also coming from traditional mass media (Zaracostas, 2020).

The term was used for the first time by Antonio Guterres, the Secretary-General of the United Nations. On March 28, 2020, he tweeted "our common enemy is #COVID19, but our enemy is also an 'infodemic' of misinformation" (Twitter, 2020). Again, UN Secretary-General, in his statement dated April 14, 2020, described the COVID-19 pandemic as the most challenging crisis the world has faced since the Second World War and defined the spread of misinformation about the disease as another epidemic. Expressing that the global "misinfo-demic" is spreading, he has invited the world to unite against the disease (UN, 2020). Also, WHO's Director-General Tedros Adhanom Ghebreyesus stated: "We're not just fighting an epidemic; we're fighting an info-demic". And he added that the spread of fake news is faster and easier than coronavirus (Naughton, 2020).

Infodemia can be defined as a contagious disease that infects the information culture (Solomon et al., 2020). In the case of a rapidly evolving situation such as the COVID-19 pandemic, it is not always easy to determine what the truth is (Eysenbach, 2020).

Personal prejudices and fears are effective in the spread of fake news (Laskar & Reyaz, 2021). In addition, life practices originating from culture and geography are other factors. For example, in countries such as India, Pakistan, Bangladesh, and Iran, homeopathy, herbal medicine, alternative medicine, and old-fashioned quackery are very common. There are circulating suggestions that eating garlic can prevent viruses, blowing hot air up the nose kills the virus, vitamin C protects people from the virus, etc. (Naeem et al., 2021). Moreover, in Iran, after the spread of rumors that alcohol prevents catching COVID-19, many people lost their lives because of drinking fake alcohol. Iranian media reported that around 300 people lost their lives due to this reason (Bag, 2020).

To combat the infodemic, 13 countries including Australia, France, and Chile have issued a statement. With the emergence of the COVID-19 virus and the declaration of a pandemic, the countries that drew attention to the increase in fake news expressed the difficulty of the situation. Emphasizing that the increase in fake news will increase violence and conflicts in society as well as the negative effects on human health and safety, they demanded that people stop spreading false information (WHO, 2020).

The struggle against false/fake news and rumors about the COVID-19 pandemic has left the states in the dilemma of protecting the freedom of speech and expression of individuals and implementing certain restrictions and prohibitions. For example, China has maintained its authoritarian management approach by restricting fake/false news about the COVID-19 pandemic with strict control. Leaving aside the principle of freedom of expression, it censored the information on social media platforms on the grounds of social order, public safety, and social responsibility (Rodrigues & Xu, 2020).

Naneem et al. (2021) recommend three strategies for winning a war against the current infodemic of misleading information about COVID-19; train people how to identify and recognize fake news stories, stop tolerating pseudoscience health practices, and swamp the landscape with accurate information. When false information emerges, its negative effect can be reduced if the correct information starts to be shared on social media by experts (Zhang et al., 2021). Also, the development of a real-time information-sharing system, drawing from data and analyses from a range of social media platforms, in multiple languages, and across the global diaspora may also be effective (Depoux et al., 2020).

Social Media and Infodemic

The popularization of social media has significantly increased the credibility of shared personal views and allowed them to spread more quickly. It has also been effective in the transparent and democratic sharing of scientific data (Orso et al., 2020). During the COVID-19 pandemic, social media, which is an important tool for information sharing, also mediated the spread of fake news.

Undoubtedly, social media is not the only source of information, there are many other sources to access information. For example, The information "Cake" model deals with information resources at four different levels. The model illustrates a four-layered wedding cake, where each layer is designated as science, policy and practice, news media, and social media. The smallest layer in the model is science and the largest layer is social media. Science stands for strict, rigorous, and selective knowledge production and is at the top. On the contrary, social media contents usually represent large-scale, uncontrolled, and unfiltered information produced by the public (Eysenbach, 2020). This leads to the production of large amounts of fake news and places social media, which has also become a news/information platform, in the middle of the discussions about getting and giving accurate news (Yegen, 2018).

People's belief in comments containing misinformation to adhere to their ideas, closing themselves to other opinions, and thinking that the comments reflect the truth play a major role in the increase in the production and spread of fake news (Kavaklı, 2019). Considering the motivations of individuals to share fake news, altruism, information seeking, information sharing, socializing, and passing time come to the fore. Self-sacrifice, i.e., caring and helping others, stands out as the most important motivation for sharing fake news (Apuke & Omar, 2021). The spread of such news, on the other hand, is influenced by personal prejudices and fears (Laskar & Reyaz, 2021).

The same words are often used in fake and real news published on social media (Patwa et al., 2021). This causes readers not to question whether the news is real or fake. In addition, fake news on social media spreads faster than true news. In a study on Twitter, it was determined that the rate of spread of fake news is 10 times faster than real news. The difference was related to the newness of the messages and the emotional reactions of the recipients (Vosoughi et al., 2018).

In the COVID-19 pandemic, disinformation in social media has emerged multidimensionally. A lot of false information has spread online, from how the virus is transmitted to unfounded treatment approaches and from messages that will create panic to conspiracy theories (Kalsın, 2020). In a study, it was determined that more than half of the fake news about COVID-19 spread through social media. It has been observed that these reports are gathered into three groups: false claims, conspiracy theories, and pseudoscientific health therapies (Naeem et al., 2021).

In a study conducted in Spain, the most reliable sources of information about the coronavirus were determined as printed newspapers and radio. Because the spread of fake news about COVID-19 on these channels is very low. They were followed by news agencies, e-newspapers, and television. On the contrary, Facebook and WhatsApp, the two main social media outlets preferred for accessing information, showed low reliability as the channels with the highest probability of producing false news (Fernández-Torres et al., 2021).

Twitter comes first among the platforms where the highest amount of false news about COVID-19 is published on social media (Soğukdere & Öztunç, 2020; Topsakal, 2021). It is known that as well as ordinary users, many politicians and scientists, who are accepted as authorities, spread untrue or unproven information via Twitter. For example; The French Minister of Health, Oliver Véran, in his tweet at the beginning of the pandemic, stated that taking anti-inflammatory drugs such as ibuprofen and cortisone would cause the disease to worsen and recommended that those with cold complaints take paracetamol. The tweet was re-tweeted 40,000 times in 3 days, but scientists could not find any scientific evidence confirming this information in their research (Orso et al., 2020).

"Fake news is taking over social media and putting public health at risk" (Naeem et al., 2021). The spread of fake news about the COVID-19 virus has the potential to affect individuals' decisions to get vaccinated (Montagni et al., 2021). Moreover, misinformations about vaccines are not taken seriously enough (Horton, 2020). But, interestingly, vaccine proponents and opponents struggle for similar demands. On the one hand, vaccine opponents, who state that the pandemic and the vaccine are biological weapons developed and spread to reduce the world population, oppose the vaccine for the welfare and future of society. On the other hand, vaccine proponents advocate vaccination also for the welfare and future of society and accuse those who are against vaccination of endangering public health (Demirhan & Başçoban, 2021).

WHO states that in the fight against COVID-19, safe and effective vaccines are of critical importance in changing the course of the pandemic and ending the disease. Underlining that it is vaccination, not vaccines, that will end the epidemic, WHO (2021) states: "We must ensure fair and equitable access to vaccines, and ensure every country receives them and can roll them out to protect their people, starting with the most vulnerable". COVID-19 vaccines should be safe for public health and have as few side effects as possible, however, confidence in the vaccine is jeopardized by disinformation (Marco-Franco et al., 2021). In addition, vaccine-related misinformation poses a serious threat not only to public health but also to national economic security (Horton, 2020).

Although vaccination against the COVID-19 virus has started in many countries today, vaccine opponents also share their views through channels suitable for manipulating public opinion (Rzymski et al., 2021). This situation causes prejudice against the vaccine in people who are exposed to negative opinions about the vaccine (Talabi et al., 2021).

The negativities that develop in individuals after vaccination are the leading cause of the spread of the anti-vaccine discourse (Demirhan & Başçoban, 2021; Larson et al., 2018). In Turkey, vaccine opponents posted on Twitter under the hashtag #aşıolmayacağım. When the contents of the tweets were analyzed, it was determined that the opposition to the vaccine was mostly based on the idea that the vaccine is harmful and/or unlicensed (Demirhan & Başçoban, 2021). In Qatar, it has been seen that false information and rumors such as instilling panic, making jokes, and suggesting unrealistic treatment methods constitute the majority of tweets about COVID-19 vaccines (Nakov et al., 2021).

Additionally, the media can guide people about vaccine reliability and public health. In a study, it was determined that news sites included news about vaccines in line with their publication policies (Topsakal & Ferik, 2021).

Fake News and Fact-Checking Websites

Attempts to verify information on the Internet have become widespread all over the world in recent years. Related initiatives discuss the accuracy of news and content and perform accuracy checks (Yegen, 2018). First Draft News, the pioneer of these initiatives started its operations in 2015. Explaining its mission of protecting communities from false and misleading information, the initiative aims to provide guidance on how to verify content on social networks. Funded by 14 organizations including Facebook Journalism Project, Google News Initiative, and Media and Democracy Fund in 2020, FDN has included many media organizations and news agencies. Another initiative that started fact-checking activities in 2017 is Google. Collaborating with 115 news verification institutions to provide this service in the relevant period, Google also started to work with the Doğruluk Payı from Turkey.

In recent years, reverse image search has become a frequently used method for fact-checking. Just like Google, TinEye is a website used for this purpose. The Canadian-based image search website quickly confirms the originality and source of any image when a URL link is entered.

There are many fact-checking platforms to verify the news. Tools such as PolitiFact, Boomlive, Newschecker, Snopes, as well as the IFCN chatbot and Google fact-check-explorer (Patwa et al., 2021) are available to that end. In Turkey, 8 websites are operated to verify the news. They can be listed as Yalansavar, Doğruluk Payı, Malumatfuruş, Teyit, Günün Yalanları, Fact-Checking Turkey, Doğruluğu Ne? (Akyüz, 2020), and Doğrula websites as of 2021. However, it's hard to say that these platforms, which take important steps to prevent disinformation in social media, have fully announced their existence and purposes to users yet (Yegen, 2018).

Turkey's first fact-checking website, Yalansavar, which started its operations in 2009, was established to "contribute to spreading the habit of critical thinking which is essential for a healthy society and scrutinizing various unfounded claims" (yalansavar.org, 2021). Doğruluk Payı, on the other hand, is a fact-checking website that was established in 2014 and is a member of the International Fact-Checking Network (IFCN) (Ünver & Edam, 2020). Since its inception in 2015, Malumatfuruş makes examinations for verification/falsification purposes focusing especially on columns as well as false news, rumors, and urban legends. Having started its activities in November 2016, Teyit aims to develop sustainable solutions to the misinformation problem by analyzing suspicious information on the internet. The website, which is a member of IFCN, classifies the content it analyzes as, "true, false, mixed, and unable to finalize". It presents the types of false information in seven different classifications as, "Misleading, imposter, manipulation, parody, fabricated, false connection, and false context" (teyit.org, 2021). Günün Yalanları, which started its activities in 2015, states that it is investigating "the news in the written and visual media, the allegations made by politicians and the accuracy of the suspicious information circulating on social media" (gununyalanlari.com, 2021). The website, which is a sister project of Fact-checking Turkey, is a Boğaziçi Global project. Fact-checking Turkey, on the other hand, is an Istanbul-based website that started operating in 2015. Verifying the news about Turkey at the international and national level, the site operates in English and states that it aims to "provide accurate information through publicly available sources" (fact-checkingturkey.com, 2021). The "Doğruluğu Ne?" website defines itself as "an impartial, independent and non-profit verification initiative that investigates questionable content on the internet and social media and aims to provide accurate information to users by verifying accuracy" (dogrulugune.org, 2021). The "Doğrula" website, which was taken over by the Digital Society Research Association as of February 2021, started its activities in 2017. A team of seven performs the verification of the claims on the website, which defines itself as "an independent initiative to combat disinformation and misinformation for the benefit of the public" (dogrula.org, 2021). The claims are evaluated by the editors according to 5 different categories: False, mostly false, half correct, mostly correct, and correct.

Interest in fact-checking websites increased with the COVID-19 pandemic in March 2020. The need to confirm the accuracy of the news regarding the pandemic has increased the use of these platforms remarkably (Siwakoti et al., 2021). These websites are usually visited by active social media users. For instance, it has been observed that active users in the USA who not only read the content but also comment on social media have a higher awareness of fake news and a higher tendency to do fact-checking (Schuetz et al., 2021).

Starting from March 11, when the first COVID-19 case was seen in Turkey, fake news also entered circulation and a lot of suspicious information started to be seen on social media. When 14 of these news are examined by the Teyit fact-checking website, one of the websites that ensure the verification of the news, it has been determined that almost all of them are false (Aydın, 2020). In another study conducted on the same website, erroneous associations and fabrications came to the fore as the most misleading elements in false news about the Coronavirus (Topsakal, 2021).

In March and April of 2020, when the COVID-19 epidemic was reported in Turkey, 92% of Doğruluğu Ne reviews, 82% of Malumatfuruş reviews, and 67% of Teyit reviews were about COVID-19. The abundance of interaction on these websites in the relevant period indicates that the demand for accurate information has increased (Akyüz, 2020, p. 432). It was determined that in the same period, 32 of the 63 suspicious social media content analyzed on the Teyit website were published on Twitter and 12 on news channels. While this indicates that Twitter is the social media outlet where fake news is published the most, it has shown that journalists can also be vulnerable to fake news (Soğukdere & Öztunç, 2020).

Methodology

The method of the study is qualitative research. According to Yıldırım and Şimşek (2013), qualitative research "uses qualitative data collection methods such as observation, interview, and document analysis and follows a qualitative process to reveal perceptions and events in a natural environment realistically and holistically" (p. 39). The data were obtained from the news about the vaccine that was analyzed in four fact-checking websites in Turkey in the second half of 2021. To determine this news, keyword research was conducted with the word "vaccine", and the contents related to COVID-19 were taken as data. For the analysis of the collected data, the content analysis technique, which is "finding how often a word is written or said according to the prepared explanatory directive" (Aziz, 2017, p. 131) was used. As Simon and Burstein stated, "content analysis can be defined as the process of coding and quantifying what people say and write according to clear instructions" (Balcı, 2011, p. 229). For this purpose, categorical data analysis was performed on the contents. According to Bilgin (2014), categorical data analysis generally means "dividing a particular message into units and then grouping these units into categories according to certain criteria. Categorization requires encoding messages, that is, processing their meaning" (p. 19). Accordingly, codes were determined for each content, and these codes were appropriately identified as themes and finally categorized.

Besides the researcher, the data obtained was also coded by a coder who is an expert in the field and the number of consensus and disagreements was determined for each question. After determining the numbers of "consensus" and "disagreement" in order to ensure the internal consistency of the themes obtained through the "agreement between coders" study, the reliability of the research was calculated using the consensus formula, which is called internal validity in the Miles and Huberman (1994) model. Miles and Huberman reliability formula:

$$\text{Reliability} = \frac{\text{Consensus}}{(\text{Consensus} + \text{Disagreement})} \times 100$$

In this formula, R: Reliability coefficient, A: Number of subjects/terms on which consensus is reached, B: Number of subjects/terms on which there is no consensus. According to the coding check, which gives internal consistency, the consensus among coders is expected to be at least 80% (Miles & Huberman, 1994; Patton, 2002). It was observed that the reliability coefficients calculated in the study varied between 80 and 100%, and the internal consistency of the study was ensured.

Sampling

In the study, the accuracy analyses of the news about the COVID-19 vaccines conducted by the fact-checking websites in Turkey, namely Doğrula, Doğruluk Payı, Malumatfuruş, and Teyit, were examined. In the selection of these websites, their detailed analyses of the claims and the presentation of sufficient evidence were effective. The study covers the 6-month period in and after July 2021.

Research Questions

What is the ratio of claims made about COVID-19 vaccines compared to other claims?
What are the results of the verification of claims about vaccines?
On which social networks are the claims usually shared?
What are the themes of fake or false claims?

Findings and Analysis

Monthly Verification Numbers on Fact-Checking Websites

A total of 1,195 analyses were performed on the fact-checking websites in the 6-month period from June to December 2021. It was determined that the fact-checking websites with the least number of analyses was

Table 1.1 Number of news analyzed by fact-checking websites and distribution by months

	Total fact-checking	*July*	*August*	*September*	*October*	*November*	*December*
Doğrula	262	37	54	51	48	36	36
Doğruluk Payı	301	49	53	49	48	50	52
Malumatfuruş	331	43	58	60	65	53	52
Teyit	301	37	54	51	58	49	52
Total	1,195	166	219	211	219	188	159

Doğrulama with 262 (21.9%) and the one with the most analyses was Malumatfuruş with 331 (27.6%). It was observed that the analyses made in the first and last months of the period were considerably low compared to other months. The number of analyses made in August and October is at the highest level (Table 1.1).

Verification Numbers for COVID-19 Vaccines on Fact-Checking Websites

Out of a total of 1,195 validations, 198 were validations related to COVID-19 vaccines. In other words, one out of every six verifications made in general is related to vaccines (16.5%). It was determined that the least verification (19.6%) regarding the vaccine was made in Malumatfuruş, and the highest verification (28.7%) was made in Teyit (Table 1.2).

Table 1.2 Fact-checking numbers for COVID-19 vaccines on fact-checking websites

	Total verification	*Related to COVID vaccines*	*Other*
Doğrula	262	50	212
Doğruluk Payı	301	52	249
Malumatfuruş	331	39	292
Teyit	301	57	244
Total	1,195	198	997

Table 1.3 Accuracy of claims about COVID-19 vaccines

	Incorrect	*Correct*	*Uncertain*
Doğrula	49	1	–
Doğruluk Payı	52	–	–
Malumatfuruş	39	–	–
Teyit	57	–	–
Total	197	1	–

Accuracy of Claims About COVID-19 Vaccines

According to the results of the fact-checking analysis of 198 claims regarding COVID-19 vaccines, only 1 (0.5%) claim was found to be true. 197 (99.5%) of the claims were untrue or false; it has been seen that some of them are fabricated, some are distorted, and some are misleading (Table 1.3).

Media Spreading Misinformation About COVID-19 Vaccines

Considering the websites where the verified claims emerged, social networks and internet sites came to the fore. It has been determined that the social network where the allegations spread the most is Twitter. A total of 158 (79.8%) claims were spread from this platform. Facebook, Instagram, Youtube, and TikTok were platforms where very few claims were shared. It has been determined that blog-style sites and news sites on the Internet are also sources of allegations. It was observed that 15 (7.57%) claims were shared on news sites and 10 (5.05%) claims were shared on websites (Table 1.4).

Table 1.4 Media spreading misinformation about COVID-19 vaccines

	Twitter	*Facebook*	*Instagram*	*Youtube*	*TikTok*	*News sites*	*Websites*
Doğrula	40	1	1	–	1	5	1
Doğruluk Payı	46	3	3	–		–	–
Malumatfuruş	38	–	–	1		–	–
Teyit	34	1	–	2	2	10	8
Total	158	5	4	3	3	15	9

Table 1.5 The locality of claims regarding COVID-19 vaccines

	Total	*Domestic*	*Abroad*
Doğrula	50	7	43
Doğruluk Payı	52	7	45
Malumatfuruş	39	3	36
Teyit	57	6	51
Total	198	23	175

The Locality of Claims Regarding COVID-19 Vaccines

When we look at whether the source of the claims about COVID-19 vaccines is local or abroad, it has been determined that the claims are mostly from abroad. Generally, in these claims, the source of the claim is also added to the image (Table 1.5).

Since the claims are in a foreign language, people who do not speak a foreign language can accept that they are true without being able to verify them. This situation can be considered as a factor in the rapid spread of fake news. In addition, it was seen that most of the news was analyzed by foreign fact-checking websites (Snopes, AFP, Reuters, etc.), and it was confirmed beforehand that it was, in fact, fake news. However, those who do not speak a foreign language or want to believe the news do not seem to need such confirmation.

Categorical Analysis of Claims About COVID-19 Vaccines

It has been observed that the verified claims regarding COVID-19 vaccines are grouped under two categories/four themes, internal and external. These themes are classified around the side effects/harms of vaccines, the vaccine content, conspiracy theories, and manipulation as codes (Table 1.6).

Vaccine Side Effects/Harms

The category with the highest number of claims is the category with claims about "vaccine side effects/harms". In this category, which includes claims about the results that will be encountered if vaccinated, claims about many side effects and harms have been shared.

Table 1.6 Categorical analysis of claims about COVID-19 vaccines

Categories	*Themes*	*Codes*
Internal	Vaccine side effects/harms	It causes sudden death
		It increases the death toll
		It does not protect, it makes you even sicker
		It causes birth defects
		It increases contagion
		It causes infertility
		It causes complications
		It causes blood clots
		It causes heart attacks
		It causes facial paralysis
		It causes stillbirth
		It destroys skills
		It causes a cytotoxic effect
		It causes GBS disease
	Vaccine content	It contains graphene oxide
		It contains fetal cells
		It contains a microchip
		It contains parasites
		It contains heart attack medication
External	Conspiracy theories	It was produced for commercial purposes
		It was developed to reduce the population
		It was developed to modify human DNA
		It had been planned long in advance
		It was developed to control people
		Anti-vaccine advocates get killed
	Manipulation	Doctors do not get vaccinated
		Rich people do not get vaccinated
		Anti-vaccine protesters are holding demonstrations
		Vaccine opponents are burning vaccine warehouses
		Vaccine opponents commit suicide to avoid getting vaccinated
		Vaccine manufacturers do not accept responsibility
		People are forced to get vaccinated
		Vaccines skipped animal trials
		Expired vaccine shots are given to people
		Vaccines are not approved

Claim: "Pfizer COVID-19 vaccine contains aborted fetal cells" (Twitter). This tweet received high interaction from many social media users. The basis of the claim is that aborted fetal cells are used as vaccine ingredients. However, there is no evidence to support this claim. Studies provide evidence that these claims about Pfizer COVID-19 vaccines are false.

Claim: "BioNTech announced that people who want to have children should not get vaccinated" (Twitter). It has been determined that some of the news shown as evidence for this claim made by a Twitter user were outdated and contained incomplete information. In addition, it is known that the BioNTech company has not made such a statement and recommends that pregnant women also get vaccinated.

Claim: "FDA says COVID-19 vaccine increases heart attack 71 times" (Twitter). A claim frequently made on social media is that vaccines increase the risk of a heart attack. The basis of the claim lies in a speech at the Vaccines and Related Biological Products Advisory Committee meeting organized by the American Food and Drug Administration (FDA). The person named Steve Kirsch, who voiced the claim at the meeting, is neither a doctor nor has any ties to the FDA. In addition, the decisions taken by this committee, in which the FDA evaluates expert recommendations, are not binding. Therefore, this claim does not reflect the truth.

Vaccine Content

Claim: "Pfizer-BioNTech vaccine contains 99% Graphene Oxide" (Twitter). According to this claim on Twitter, the vaccine contains graphene oxide and this content affects the human body. Graphene oxide, which is a 2D nanomaterial composed of highly oxidized carbon atoms, is used in many technological fields like solar cells, biomedical research, etc. This claim, which is based on a Spanish article, does not reflect the truth. Because the article is not included in official and reliable academic sources. In addition, it has been announced by the American Food and Drug Administration that there is no graphene oxide among the substances in the BioNTech vaccine.

Claim: "Vaccines contain parasites". On Twitter and other social media platforms, it has been claimed that the vaccines contain a parasite that is used to hijack the human brain. However, there is no organism/parasite as claimed in the vaccine contents announced by the vaccine companies over the internet. The statements of the subject matter experts reveal that this information is fabricated. In addition, according to the results of the fact-checking, the image shared and claimed to be the parasite belongs to a sea creature, the crinoid, and was taken from a video uploaded to Youtube in 2019.

Claim: "Pfizer secretly added heart attack drugs to children's Covid vaccines" (Twitter). The website claimed that Pfizer has added *tromethamine*, a chemical that is used to treat heart attacks, to Covid vaccines for children. According to this claim, which later spread on other social media networks, this drug is used to eliminate the risk of heart attack due to vaccination. However, the drug is not used in the treatment of heart attacks. According to a statement from Pfizer officials to the verification agency AFP, *tromethamine* is added to extend the shelf life of vaccines. The basis of the claims is that this drug has been added to the vaccine at a later period. However, Pfizer stated that speed was given importance in the production of the first vaccines, but increasing the durability of vaccines has gained growing importance in the current stage.

Conspiracy Theories

Claim: "Sweden has put 'identifying the vaccine passport with a microchip implanted under the skin' into practice" (Twitter). In the video that is the source of this claim on Twitter, it is only discussed whether such a practice is viable. Therefore, the claim is unfounded and no microchip was placed under the skin through vaccines.

Claim: "Anti-vaccine leaders are mysteriously killed" (Twitter). In a claim published on Twitter, it was claimed that vaccine opponents either died or disappeared. These allegations were voiced both in Turkey and abroad. It has been claimed that in Turkey, Cardiology Professor Canan Karatay did not appear on (was banned from) television after the claims she made about the Coronavirus on television screens. Similarly, it has been claimed that African leaders were killed because of their anti-vaccine statements. However, some of these leaders do not even have any anti-vaccine statements. Looking at the causes of death of these people, it is known that Tanzanian President John Magulfi and Burundi Prime Minister Pierre

Nkurinziza died due to heart problems. The Prime Minister of Ivory Coast, Hamed Bakayoko died because of cancer. Also, no evidence was found that Eswatini Prime Minister Ambrose Dlamini, who died two weeks after testing positive for COVID-19, was killed. There is also information that Dlamini strictly implemented measures against the disease in his country. As a result, it is seen that contrary to the conspiracy theories, the deceased leaders died naturally.

Claim: "Vaccines make people 'humanoids' and deprive them of their human rights" (Twitter). In a tweet on Twitter, it was claimed that vaccines would cause changes in the human genome and thus people would lose all their rights. Furthermore, it has been suggested that studies to show that those who are vaccinated cannot be counted as human have been started in the USA. However, as a result of studies on mRNA vaccines, it has been determined that these vaccines do not cause any temporary or permanent changes in the human genome and do not turn people into another species. Therefore, the allegations do not reflect the truth. In addition, as a result of the fact-checking, it was determined that the tweet owner, who came up with the conspiracy theory, used a fake academic identity.

Manipulation

The category of manipulative claims about the vaccine is the second category with the highest number of claims. Claims that the doctors who developed the vaccine, Bill Gates, the owner of Microsoft, one of the world's leading companies, as well as the wealthy and vaccine manufacturers, did not get vaccinated are included in this group. In addition, the alleged protest demonstrations by anti-vaccine advocates also belong to this theme.

Claim: "The founders of BioNTech, Özlem Türeci and Uğur Şahin, have not got a Covid-19 vaccine shot" (Twitter). The claim does not reflect the truth. The reason why two scientists were not vaccinated at the first stage is that they are not among the priority groups to be vaccinated. Şahin stated that he received his first dose of the vaccine together with BioNTech employees in an interview he gave on March 19, 2021 (https://archive.is/NdEPx).

Claim: "Police inflict violence against those who did not get vaccinated in France" (Twitter). The video at the center of the claims was posted on

Twitter. However, the footage has nothing to do with anti-vaccination. The related video is dated November 28, 2020, and belongs to a protest against police violence.

Claim: "Anti-vaccine rally held in Vienna" (Twitter). The photograph used in the announcement, which is claimed to belong to the protest of anti-vaccine protesters in Vienna, does not reflect the truth. The related photo was originally shot on March 10, 1991, in Manezh Square in Moscow, the capital of Russia, during the collapse of the USSR.

Claim: "Maltepe District Governor was exiled for not allowing the anti-vaccine rally". This claim has been published on many news sites. However, it has been determined that the decision regarding the appointment of the district governor to the new place of duty was taken approximately one month before this date. The Ministry of Interior also confirmed this information. As a result, it is true that the Maltepe District Governorate did not allow the rally, but the information that it was expelled for this reason is a lie.

Results

In the study, the fact-checkings about the COVID-19 vaccines on the fact-checking websites in Turkey were analyzed. The number of fact-checking news about the vaccine among all fact-checking was 198 (16.5%). This number does not include news about the COVID-19 virus, but only news about the vaccine. Of the 198 analyzed items, only 1 is correct and 197 is incorrect. These contents were shared on social media networks and sites on the internet. 158 (79.7%) of the contents were shared on the social media network Twitter and 15 (7.5%) on news websites. Considering the locality of the shared contents, it was determined that 23 (11.7%) contents originated from Turkey and 175 (88.3%) contents originated from abroad.

In the content analysis for the content on the fact-checking websites, two categories, four themes, and 28 codes were found. Categories were determined as internal and external. The themes in the internal category were classified as the side effects/harms of the vaccine and the content of the vaccine while the themes in the external category were classified as conspiracy theories and manipulation. Codes such as causing sudden death, infertility, heart attack, and facial paralysis came to the fore in the theme of side effects/harms of the vaccine; and the presence of graphene oxide, fetus, microchip, and parasite came to the fore in the theme of vaccine content. Vaccine conspiracy theories cover the codes that the vaccine was produced for commercial purposes, developed to reduce the population, that Anti-vaccine advocates get killed, etc. while the theme of manipulation covers the codes that rich people and vaccine manufacturers are not vaccinated, anti-vaccine advocates are holding demonstrations, vaccines are not approved, etc.

Conclusion

This study attempted to determine whether the contents related to COVID-19 vaccines were true or not and what kind of themes they consisted of. It has been suggested that vaccines increase mortality rates, cause many side effects, and contain substances intended to control humanity. It has been stated that vaccine opponents have been eliminated, and those who produced the vaccine, along with rich people, have not gotten vaccinated. While the importance of vaccination for public health and ending the pandemic is obvious, the spread of such false news is extremely dangerous. In order to prevent this, it is necessary to carry out studies particularly on social media networks. Although companies such as Facebook, Twitter, and Google accelerated their efforts to combat misinformation in the middle of the pandemic, a multifaceted and large-scale intervention still has not been made (Zhang et al., 2021). Social media companies such as Facebook and Twitter should eliminate misinformation about COVID-19 vaccines. Trusted politicians and public figures should give speeches to encourage vaccination. Scientists working on vaccines should raise the standards of their work and work as independently as possible from their sponsoring manufacturers. Journalists should not mediate the spread of misinformation. Lawmakers should make legal arrangements to address the source of misinformation (O'Connor & Weatherall, 2019). Knowing that they will face a penalty if they post fake news can stop them. Another measure can be to increase the number of correct information. Future studies may attempt to determine why those who share such news are against vaccination and to learn the thoughts of those who support them. Thus, it could be possible to get to the root of the problem. In addition, the study on the content of the fact-checking websites in Turkey can be carried out in other countries to reveal the similarities or differences. The effect of cultural and geographical differences on the spread of fake news can be investigated.

References

Akyüz, S. S. (2020). Yanlış Bilgi Salgını: COVID-19 Salgını Döneminde Türkiye'de Dolaşıma Giren Sahte Haberler. *Akdeniz Üniversitesi İletişim Fakültesi Dergisi, 34*, 422–444.

Apuke, O. D., & Omar, B. (2021). Fake news and COVID-19: Modelling the predictors of fake news sharing among social media users. *Telematics and Informatics, 56*, 101475.

Aydın, A. F. (2020). Post-Truth Dönemde Sosyal Medyada Dezenformasyon: Covid-19 (Yeni Koronavirüs) Pandemi Süreci. *Asya Studies-Academic Social Studies/Akademik Sosyal Araştırmalar, 4*(12, Summer), 76–90.

Aziz, A. (2017). *Sosyal Bilimlerde Araştırma Yöntemleri ve Teknikleri* (11. Basım). Nobel Yayıncılık.

Bag, M. (2020). İran'da Alkolün Koronavirüsten Koruduğu Söylentileri Nedeniyle Sahte İçki İçen Yüzlerce Kişi Öldü. *EuroNews*. https://Tr.Euronews.Com/2020/03/27/İran-Da-Alkolun-Koronavirusten-Korudugu-Soylentileri-Nedeniyle-Sahte-İcki-İcen-Yuzlerce-Ki. Accessed 2 December 2021.

Balcı, A. (2011). *Sosyal Bilimlerde Araştırma Yöntem, Teknik ve İlkeler* (9. Baskı). Pegem Akademi.

Bilgin, N. (2014). *Sosyal Bilimlerde İçerik Analizi*. Siyasal Dağıtım.

Buchanan, M. (2020). Managing the infodemic. *Nature Physics, 16*, 894. https://doi.org/10.1038/S41567-020-01039-5. Accessed 1 December 2021.

Demirhan, K., & Başçoban, A. E. (2021). Covid-19 Salgınında Aşı Karşıtlarının Twitter'da# Aşıolmayacağım Etiketi Altında Yaptığı Paylaşımların Kamu Sağlığı Politikaları Bağlamında Analizi. *Yeni Medya, 11*, 102–115.

Depoux, A., Martin, S., Karafillakis, E., Preet, R., Wilder-Smith, A., & Larson, H. (2020, April). The pandemic of social media panic travels faster than the COVID-19 outbreak. *Journal of Travel Medicine, 27*(3).

Eysenbach, G. (2020). How to fight an infodemic: The four pillars of infodemic management. *Journal of Medical Internet Research, 22*(6), E21820.

Fernández-Torres, M. J., Almansa-Martínez, A., & Chamizo-Sánchez, R. (2021). Infodemic and fake news in Spain during the COVID-19 pandemic. *International Journal of Environmental Research and Public Health, 18*(4), 1781.

Horton, R. (2020). Offline: Managing the COVID-19 vaccine infodemic. *Lancet, 396*(10261), 1474. https://onu.delegfrance.org/IMG/pdf/cross-regional_statement_on_infodemic_final_with_all_endorsements.pdf. Accessed 29 November 2021.

JHU. (2022). *Global Map for Covid-19.* https://coronavirus.jhu.edu/map.html. Accessed 1 February 2022.

Kalsın, B. (2020). *Kovid-19'dan Kaçarken Dezenformasyona Tutulmak.* https://www.Aa.Com.Tr/Tr/Analiz/Kovid-19dan-Kacarken-Dezenformasyona-Tutulmak/1781644. Accessed 30 November 2021.

Kavaklı, N. (2019). Yalan Haberle Mücadele ve İnternet Teyit/Doğrulama Platformları. *Erciyes İletişim Dergisi, 6*(1), 663–682.

Larson, H., De Figueiredo, A., Karafillakis, E., & Rawal, M. (2018). *State of Vaccine Confidence in the EU 2018.* https://Ec.Europa.Eu/Health/Sites/Default/Files/Vaccination/Docs/2018_Vaccine_Confidence_En.Pdf. Accessed 15 November 2021.

Laskar, K. A., & Reyaz, M. (2021). Mapping the fake news infodemic amidst the COVID-19 pandemic: A study of Indian fact-checking websites. *Journal of Arab & Muslim Media Research, 14*(1), 93–116.

Marco-Franco, J. E., Pita-Barros, P., Vivas-Orts, D., González-De-Julián, S., & Vivas-Consuelo, D. (2021). COVID-19, fake news, and vaccines: Should regulation be implemented? *International Journal of Environmental Research and Public Health, 18*(2), 744.

Miles, M. B., & Huberman, A. M. (1994). *Qualitative data analysis: An expanded sourcebook* (2nd ed.). Sage.

Montagni, I., Ouazzani-Touhami, K., Mebarki, A., Texier, N., Schück, S., Tzourio, C., & Confins Group. (2021). Acceptance of a Covid-19 vaccine is associated with ability to detect fake news and health literacy. *Journal of Public Health, 43*(4), 695–702.

Naeem, S. B., Bhatti, R., & Khan, A. (2021). An exploration of how fake news is taking over social media and putting public health at risk. *Health Information & Libraries Journal, 38*(2), 143–149.

Nakov, P., Alam, F., Shaar, S., Martino, G. D. S., & Zhang, Y. (2021). A second pandemic? Analysis of fake news about COVID-19 vaccines in Qatar. *Arxiv Preprint Arxiv: 2109.11372.*

Naughton, J. (2020). Fake news about Covid-19 can be as dangerous as the virus. *The Guardian.* https://www.theguardian.com/commentisfree/2020/mar/14/fake-news-about-covid-19-can-be-as-dangerous-as-the-virus. Accessed 26 February 2022.

O'Connor, C., & Weatherall, J. O. (2019). *The misinformation age.* Yale University Press.

Orso, D., Federici, N., Copetti, R., Vetrugno, L., & Bove, T. (2020). Infodemic and the spread of fake news in the COVID-19-era. *European Journal of Emergency Medicine, 27*(5), 327–328.

Öztürk, A. (2021). COVID-19 Pandemi Sürecinde Bilişim Teknolojileri Bağımlılığı. *Ağrı İbrahim Çeçen Üniversitesi Sosyal Bilimler Enstitüsü Dergisi, 7*(1), 195–219.

Patton, M. Q. (2002). *Qualitative research and evaluation methods* (3rd ed.). Sage Publications Inc.

Patwa, P., Sharma, S., Pykl, S., Guptha, V., Kumari, G., Akhtar, M. S., Ekbal, A., Das, A., & Chakraborty, T. (2021, February). Fighting an İnfodemic: Covid-19 fake news dataset. In *International workshop on combating on line hostile posts in regional languages during emergency situation* (pp. 21–29). Springer.

Rodrigues, U. M., & Xu, J. (2020). Regulation of COVID-19 fake news İnfodemic in China and India. *Media International Australia, 177*(1), 125–131.

Rzymski, P., Borkowski, L., Drąg, M., Flisiak, R., Jemielity, J., Krajewski, J., Mastalerz-Migas, A., Matyja, A., Pyrć, K., Simon, K., Sutkowski, M., Wysocki, J., Zajkowska, J., & Fal, A. (2021). The strategies to support the COVID-19 vaccination with evidence-based communication and tackling misinformation. *Vaccines, 9*(2), 109.

Schuetz, S. W., Sykes, T. A., & Venkatesh, V. (2021). Combating COVID-19 fake news on social media through fact checking: Antecedents and consequences. *European Journal of Information Systems*, 1–13.

Siwakoti, S., Yadav, K., Bariletto, N., Zanotti, L., Erdoğdu, U., & Shapiro, J. N. (2021). How COVID drove the evolution of fact-checking. *Harvard Kennedy School (HKS) Misinformation Review.*

Soğukdere, Ş., & Öztunç, M. (2020). Sosyal Medyada Koronavirüs Dezenformasyonu. *Kastamonu İletişim Araştırmaları Dergisi, 5*, 59–85.

Solomon, D. H., Bucala, R., Kaplan, M. J., & Nigrovic, P. A. (2020, November). The "infodemic" of COVID-19. *Arthritis & Rheumatology, 72*(11), 1806–1808.

Talabi, F. O., Ugbor, I. P., Talabi, M. J., Ugwuoke, J. C., Oloyede, D., Aiyesimoju, A. B., & Ikechukwu-Ilomuanya, A. B. (2021). Effect of a social media-based counselling intervention in countering fake news on COVID-19 vaccine in Nigeria. *Health Promotion International.*

Topsakal, T. (2021). Dijital Ortamda Yanlış Bilgi ve Haberlerin Yayılması: Koronavirüs Salgın Haberlerine Dair Bir İnceleme. *İnönü Üniversitesi İletişim Fakültesi Elektronik Dergisi (İnif E-Dergi), 6*(1), 383–401.

Topsakal, T., & Ferik, F. (2021). Haber Sitelerinin Covid-19 Aşılarına Yönelik Yaklaşımı Ve Haber İçeriklerinin Değerlendirilmesi. *Akdeniz Üniversitesi İletişim Fakültesi Dergisi, 35*, 370–386.

Twitter. (2020). https://twitter.com/who/status/1228683949796470784
UN. (2020). *The Secretary-General Message on COVID-19 and Misinformation*. https://Www.Unodc.Org/Unodc/Press/Releases/2020/April/Message-On-Covid-19-And-Misinformation.Html. Accessed 29 November 2021.
Ünver, H. A., & Edam, O. C. (2020). *Türkiye'de Doğruluk Kontrolü ve Doğrulama Kuruluşları*. Centre for Economics and Foreign Policy Studies.
Vosoughi, S., Roy, D., & Aral, S. (2018). The spread of true and false news online. *Science, 359*(6380), 1146–1151.
WHO. (2020). *Cross-regional statement on "infodemic" in the context of COVID-19*.
WHO. (2021). https://www.who.int/emergencies/diseases/novel-coronavirus-2019/covid-19-vaccines. Accessed 2 December 2021.
Yegen, C. (2018). Doğru Haber Alma Hakkı ve Sosyal Medya Dezenformasyonunu Doğruluk Payı ve Yalansavar ile Tartışmak. *Erciyes İletişim Dergisi, 5*(4), 101–121.
Yıldırım, A., & Şimşek, H. (2013). *Sosyal Bilimlerde Nitel Araştırma Yöntemleri* (9. Baskı). Seçkin Yayınları.
Zaracostas, J. (2020). *How to fight an infodemic. WHO's newly launched platform aims to combat misinformation around COVID-19* (World Report).
Zhang, J., Featherstone, J. D., Calabrese, C., & Wojcieszak, M. (2021). Effects of fact-checking social media vaccine misinformation on attitudes toward vaccines. *Preventive Medicine, 145*, 106408.

CHAPTER 2

Sustainable Branding in Digital Environment in Terms of Corporate Communication

Mikail Batu and Özlem Kutlu

Introduction

The purchasing behavior of customers, who can instantly reach the information in different countries and in societies with different languages, religions, and cultures can vary, thanks to information and communication technologies. This also makes it significant to record the information of customers; on the other hand, this makes it difficult for corporations to make strategic decisions about customers. The information of customers using Internet becomes an excessively valuable product for the marketing strategies of corporations. Accordingly, maintaining the existence of the corporations in the digital environment, which is the changing form of trade, depends on their branding, which is one of the implementation areas of corporate communication, and exhibiting a successful image. While the images of brands with a negative image can negatively affect

M. Batu (✉) · Ö. Kutlu
Department of Public Relations and Publicity, Ege University, Bornova, Turkey
e-mail: mikail.batu@ege.edu.tr

S. R. Mondal et al. (eds.), *New Normal in Digital Enterprises*,
https://doi.org/10.1007/978-981-19-8618-5_2

the purchasing behavior of the customers, a brand with a positive image positively affects the decision-making process of the customers (Batu & Kutlu, 2020: 56). The important thing at this point is how the brand is represented and how the process in the virtual environment is managed. This management process is related to the professional implementation of corporate communication practices that include all internal and external communication activities of the corporation (Bat, 2012). For example, if the brand value of the product or service is expected to increase in the digital environment, different implementation areas such as image, reputation, and identity should be taken into consideration with digital media strategies.

In a free competition environment, corporations are in the race to become prominent in the market where many corporations exist and products and services are similar. The main conditions for corporations to exist in this race are to make a profit by selling and to be preferred by customers who are within unlimited alternatives. In line with these goals, one of the principles that corporations make great efforts and investments in is the brand factor (Fidan & Summak, 2001: 29). The brand, which serves to distinguish a product from other products on the shelves, is the name, sign, symbol, or a combination of these that determines the identity of goods and services, enables the product to be distinguished from its competitors, facilitates communication with customers as the focal point of marketing and advertising management implementations, and differentiates the product (Aktuğlu, 2004: 16). The notion of brand with several definitions in the literature is according to the definition of the American Marketing Association; "It is the name, term, sign, symbol, design or various combinations of all of these that are used to determine the goods and services of a corporation and to diversify it from the goods and services of its competitors" (Stanton, 1975: 214). David Aaker (1991), who defines the brand as a name, sign, and symbol that diversifies corporations from each other, does not solely associate the brand with a product, but he utilizes it in a way to include the entire corporation which is the source of the product (Hague & Jackson, 1994). To Konecnik and Gartner (2007), a brand is a legal tool, which creates high value for a corporation, and an abstract value that determines the identity, logo, image, and relationships of the corporation. On the other hand, Chernatony and Riley (1998) handled the definition of the brand from a broad perspective and categorized it into 12 themes. These themes are listed below:

- Brand as a legal tool,
- Brand as a logo,
- Brand as a corporation,
- Brand as a shortcut tool (rapid information to remind in memory),
- Brand as a risk reducer,
- Brand as an identity system,
- Brand as the image in the mind of the customer,
- Brand as a value system,
- Brand as a personality,
- Brand as established relationship,
- Brand as added value,
- Brand as a developing entity.

A brand is a name, symbol, and sign that will make a product or service to be distinguished from its competitors in the eyes of the target group. To Teemu (2009), the brand, which is the totality of perceptions formed in the mind of the customer by the sum of material and moral elements related to the product and service, facilitates, shortens, and simplifies (physical and mental) the decision-making process of the customer among alternatives (Aaker & Keller, 1990: 27), and decreases product search costs. Customers with information about a specific brand can make assumptions about the features of the product that they do not know and create reasonable expectations (Keller, 2003: 6). In current marketing, branding is a process to define a product or service, to diversify it from its competitors and to create its uniqueness. In practice, branding, which means more than that, is part of the process of how customers perceive the product or corporation and become sensitive to all messages given (Yadin, 2002).

The notion of the brand, which enables the product to be perceived at an emotional, associative, and relational level as well as its physical features, dates back a long time. It is believed that numerous "handprints" indicating that a brand is a form of sign of belonging traces to 15,000 B.C. It is known that ancient civilizations such as Egyptians, Greeks, Romans, and Chinese put seals on their pottery and other items to indicate quality (Perry & Wisnom, 2003: 11). In ancient times, symbolic and decorative figures were used on products as emblems to represent power and authority by kings, emperors, and governments. Japanese used chrysanthemum, Romans used eagle, and the French used the lion as symbols (Knapp, 2000: 87). In European history, the brand began with medieval guilds putting commercial signs on their products to protect themselves and their customers against low-quality products, and with artists who signed their works in fine arts (Kapferer, 2008).

The United Kingdom, which realizes signs and symbols on branded products facilitate tax collection, enacted the "Bakers Marking Law" in 1266. Thus, bakers showed the first examples of trademarks in the modern sense by labeling their breads that indicated where they were produced (Perry & Wisnom, 2003: 11–12). The notion of the brand, which has been used from those dates until today, has been subject to many definitions, has been attributed different meanings, and it has been a phenomenon that serious investments and efforts has been made in order to protect the existence of corporations in a free competitive market.

Brand perception in the mind of the customer has become an essential element of the marketing mix for corporations as it directs the purchasing behavior of the customer.

Nowadays, the investments made by corporations for a strong brand and the positive brand perception in the mind of the customer lead the corporations to gain a stronger sustainable competitive advantage compared to their competitors (Pappu et al., 2005: 143).

Sustainable Brand and Brand Leadership in Digital Environment

The notion of the brand, which simplifies the time-consuming search and comparison process before customers make a purchasing decision, becomes significant in the digital environment as well as in traditional environments. It is observed that brands are known more in a short time in the digital environment in which information is easily accessible and information loading is excessively fast and enables customers to save time and cost. The importance of branding has gradually increased by showing continuity and gaining customer loyalty in online markets which acts and changes are measured in seconds (Rowley, 2004: 132). The emergence of Internet technology, particularly real-time interaction and the effects of the existence of many corporations in the market, made branding more complex and dynamic. For this reason, most corporations in the digital environment seek new internet brand strategies that can help them make a difference from their competitors while they try to protect their own customers (Simmons, 2007: 544).

A brand is traditionally thought to evoke a specific personality, presence, and product or service performance in the mind of the customer. In addition to providing added value, a brand simplifies the comparison process of customers in the pre-purchase process (Simmons, 2007: 544). In the digital environment, the brand is matched with experience (Rowley, 2004: 132). Thanks to the internet technology, corporations can interact with customers and change their presentations to meet individual customer needs. Unlike traditional media, internet enables a two-way communication and helps them establish a permanent relationship with customers. The customer, who has a positive experience in the digital environment, can create loyalty towards the brand. This provides an opportunity for all corporations aiming for sustainable profit flow to brand in the digital environment (Neelotpaul, 2010: 2).

"Leadership claim" refers to a corporation's claim to superiority in any subject. The ability of a brand to show a range of functional or emotional aspects and the sense of satisfaction it provides to customers is the sign of the value created by the brand. A corporation, which desires to strengthen its leadership claim, can use this value. At the same time, the size of the corporation in market conditions indicates its corporate strength and leadership. For example, a corporation may claim to have the largest market share compared to its competitors or to be the strongest leader financially (Virtsonis & March, 2009: 564). The brand value of a corporation that is a leader in its sector also has tangible value independent of its products and can be subject to purchase and sale by being on the list of the tangible assets of the corporation (Kotler, 2000: 405). In this case, the share value of the corporation increases and an increase is accomplished on the balance sheet and financial values (Fidan & Summak, 2001: 31). The leadership of a brand can also be determined by taking into account internal factors such as the "product characteristics" of the corporation or external factors such as "industry and market conditions" (Conger & Kanungo, 1987). On the other hand, brand leadership reflects the concrete competitive advantage of a corporation over other brands. In addition, leadership highlights the perceived competitive relationship between the leading corporation and other brands that follow it (Chang & Ko, 2014: 65). In this competitive relationship environment, leading brands are taken as an example and imitated by other corporations in terms of visible and invisible features. Leading corporations can also positively affect other corporations that imitate their present products or take invisible parts such as their vision and goals as an example (Chang & Ko, 2014: 64).

It is recommended that corporations use organizational structures, processes, and culture in the context of global branding for brand leadership which is regarded as the ability of a brand to accomplish constant excellence (Aaker, 1996). To Miller and Mills (2012), the more innovative, creative, and unique a brand is, the more the customer sees the brand as a leader and perceives it as a market leader. In his study, Measuring Brand Equity across Products and Markets in 1996, Aaker attributes brand leadership to the reflection of a corporation's competitive advantage over other brands and specific actions. Accordingly, whether the corporation is regarded as a leading brand, whether it has popularity, and whether it is innovative and pioneer in products and services among the questions in the scale used in the same study to determine a leading corporation.

Baek and King (2011) mentioned the factor that helps to increase the perceived leadership of a brand is the customer's trust in the brand and the information obtained by the customers to build trust. The quality of the products presented for consumption by the corporations is generally not easily observable by the buyers before purchasing but it completely appears after the purchase. Uncertainty about quality cannot be resolved before purchasing due to this feature, and therefore product quality is mentioned as an unobservable feature (Akshay et al., 1999: 259). Customers under uncertainty tend to seek more information about product or service quality before making a decision (Baek & King, 2011: 262). In today's world, where digital shopping grows and it is preferred because of its many benefits, customers can easily obtain the desired information about quality or innovation with communication tools such as mobile phones, thanks to the advancement in information technologies. Rich information obtained through additional communication applications such as search engines or daily communication also helps to increase the credibility and perceived leadership of a brand (Chang & Ko, 2014: 65).

Sustainable Competitive Advantage and Strategies to Increase Market Share

Corporations to be successful in an intensely competitive environment depend on them analyzing their customers better than their competitors. Corporations aim to accomplish a superior competitive position in the long term by having superior performance compared to their competitors (Engin, 2005: 30). The ability of corporations to have a competitive advantage is also related to create more value for their customers than competitors. More value leads the corporation to make its products or services at a lower cost or better than its competitors (Porter, 1985: 3). Depending on these issues, there are general competitive strategies that corporations can use to gain a competitive advantage. These strategies, which are also called strategies to increase market share or maintain the current market position, are cost leadership, diversification, and focus strategies (Porter, 2000: 44).

Cost Leadership Strategy

The cost leadership strategy is for a corporation to gain a competitive advantage by reducing its costs lower than all its competitors. The cost leadership strategy can only be implemented if product diversification is limited and customers seek specific common features in the product.

Diversification Strategy

Diversification strategy is to diversify the product or service offered by the corporation and become unique in the whole sector (Göl, 1995: 33). Nowadays, customers are interested in distinguishing the benefit of a brand that will work. As long as a brand does not offer a clear benefit, it does not mean much to the customer (Knapp, 2000). At this point, it is possible to identify and meet the expectations of the customers properly and to ensure that the customer is satisfied with the product and it is perceived at a higher value than the value of the product through diversification (Engin, 2005: 46). In today's conditions that change is extremely rapid, having different demands, needs, and requirements can move the diversification strategy ahead of the cost leadership strategy (Porter, 2000: 44). Diversification can be accomplished in elements such as a unique product-service, design or brand name, technology, customer service, service and support services, and brand image (Göl, 1995: 39).

Focus Strategy

Focus strategy is the strategy that corporations implement by focusing on customer groups with specific needs, in other words, "narrow target groups", instead of implementing broad-based strategies, and therefore provides a sustainable competitive advantage.

Creating a successful corporation and a successful brand requires considering customer needs and creating new needs. It is not necessary to make a new invention to create a new need for customers. It may also be adequate to develop a part of the current need and make it a category in itself. Generally, such needs are always taken into consideration, but the emergence and diversification of this need by a corporation can bring about market leadership in its sector (İlgüner, 2006: 67). Diversification of a corporation in the eyes of the customers can bring about not only the advantage of pricing strategy but also "market leadership" (Alwi & Silva, 2007: 218). To Porter (2000), the corporation gains the following advantages and can become a leader in its market thanks to the diversification strategy.

- Since customers will have limited options for product/service diversification, their sensitivity to price is lower, and this reduces the bargaining power of customers.
- As diversification leads to high-profit margins, it gives corporates more flexibility in coping with increases in input costs and corporates can reflect the increase in input costs to their customers through prices.
- Customer loyalty, which is a significant obstacle to be overcome by new corporates, makes it difficult to get into the market. Product/service diversification creates an opportunity to overcome this obstacle.
- Corporates can be protected against competition with product diversification. The corporate, which cannot completely isolate itself against the competition, does not have to compete with rival corporates thanks to its customers who are less sensitive to price.
- Corporates are in a good position against alternative products thanks to their brand loyalty.

Brand leadership for a corporation can be explained by factors such as its competitive advantage, its market share, brand value, the increase on its balance sheet and financial values, whether it has popularity, and whether it is an innovative and pioneer in products and services. In addition to all these elements, corporations need to position their brands correctly and have a combination of communication efficiency in order to reach leadership (Chang & Ko, 2014: 65). It can be regarded as a prerequisite for corporations, which accomplish branding to be ahead of their competitors in the competitive process, to create customer loyalty and to fulfill the leadership criteria. For this reason, branding, which is highly significant for corporations, means much more than communication strategies and the messages they convey (Andrei & Talabă, 2010: 311).

Sustainable Brand Image in Digital Environment

A strong corporate brand image not only creates a competitive advantage for a corporation but also creates customer loyalty and therefore, encourages the customer to purchase again (Silva & Alwi, 2008: 120). While having a positive brand image affects the decision-making process of customers, a negative image can negatively affect these decisions and behaviors (Alwi & Silva, 2007: 218). A positive corporate image helps corporations to reach a high level of performance and increases sales and corporate profitability without paying large amounts of money for advertisements (Silva & Alwi, 2008: 119).

The image of a corporation or a physical store is associated with the quality of that corporation. It is observed that corporate image, which is highly important for corporations, is also valid for corporate brands that exist in the digital environment. It has been seen in the studies that the website features of digital corporations have a positive contribution to their corporate image. In this context, in the study conducted by Kuzic et al. in 2010; "The easy connection of the websites, the content of the websites and the first impressions on the customers" are mentioned as factors that affect the corporate brand image in the digital environment. Also in the context of online shopping; Factors such as easy connection of websites, security of websites, ease to interact, and being entertainment have been identified as driving elements that affect attitudes towards image (Merrilees & Fry, 2002).

Corporate websites for corporations and brands can combine many functions such as information and image-creating strategies, direct and indirect sales functions in addition to products and services. There is no need to have additional time or to purchase domain in media tools with additional messages added to web pages. Websites that offer a unique opportunity to combine multiple types of messages to multiple stakeholders (Hwang et al., 2003) indicate that customers can evaluate a corporate brand not only cognitively but also emotionally. As long as the websites provide an acceptably better connection, the customer can create a positive evaluation and positive perception (emotional) towards the corporate brand. Website revisions with details such as keeping it updated (in order of importance), providing high-quality content with valuable information, ease of use, fast connection and fast download, and providing personalized services determine the intention of customers to visit these websites again (Supphellen & Nysveen, 2001: 2). These factors contribute positively to the images of the corporations in the digital environment and affect the customers to have positive attitudes towards them and to create brand loyalty.

Silva and Alwi (2008) suggested that the website of the corporation should be considered as a tool to strengthen the brand image, and in their study in 2007, they defined the dimensions of the corporate brand image in the digital environment; "Agreeableness", "Innovation", "Competence", "Aesthetic and Visual Features", and "Flexibility" (Alwi & Silva, 2007: 237). In the study conducted by Silva and Alwi in 2008; the corporate brand image components of Davies et al. (2003) in the physical environment were utilized by adapting them to the digital environment. These factor groups and variables are shown in the Table 2.1.

Successful Branding in Digital Environment

With new technologies, market trends shift the balance of power from corporations to customers. The possibility of real and simultaneous interaction with the Internet and the emergence of an independent and unlimited number of market areas lead corporations to move their branding strategies to the digital environment. Branding in the digital environment is defined as the process of creating value (Ibeh et al., 2005: 355). To Simmons (2007), it is significant to fulfill the following criteria for successful branding in the digital environment.

- To create an online brand that acts as fast as possible to be a pioneer in the market,

Table 2.1 The corporate brand image components of websites

Agreeableness	*Honest, *Sincere, *Unreserved and coherent, *Always ready and appropriate, *Reliable, *Warranting and comforting, *Agreeable, *Socially responsible, *Friendly, *Entertaining, *Supporting and helpful (Technology, services etc.), *Concerned
Enterprise	*Courageous, *Follower of the latest developments and technology, *Thrilled, *Cool-blooded, *Dynamic, energetic, young, *Extroverted, *Creative and imaginative, *Keeping the website up to date, *Having an innovative spirit
Competence	*Strong in terms of security, *Solid and strong in terms of technical issues, *Having a corporate structure, *Success-oriented, *Ambitious and determined, *Believable and trustworthy (technological infrastructure and services, etc.), *Hardworking, *Having a guiding spirit
Strict RULES	*Selfish, *Authoritative, *Controlling, *Arrogant, *Aggressive, *Introverted
Aesthetic and visual features	*Cute and pleasant, *Original, prominent and privileged, *Courtesy, *Showy and fashionable, *Prestigious, *Beautiful and simple, *Perky, *Elite
Flexibility	* Ordinary, randomly and shallowly, *Plain and simple, *Tolerant
Unpleasant attitudes	*Rude and irregular, *Strict, *Masculine

Source Allport as cited in Alwi, 2009

- To get information about target customers with a systematic understanding,
- To carry out an advanced "order" and "return" process by going beyond to raise awareness for websites, and therefore building trust and relationships,
- To establish stronger relationships with target customers through unique messages, unique functionality, and unique personalization techniques,
- To provide a quality product and service within a unique positioning and communication concept,
- To ensure that brand promises are consistently kept.

To Rowley (2004), a successful brand has three aspects.

- A brand depends on customer perception.

- Perception is affected by the added value of the product and service.
- The added value should be sustainable.

Branding in the literature focuses on the criteria of understanding customers, communicating with them, and maintaining interaction. Simmons (2007) examines branding in the digital environment on the following four criteria by adding "content" to these three criteria.

Customer Understanding

The basic building block of successful branding in the digital environment is the need to understand online customers (McQuade et al., 1996). Understanding customers contributes to the development of trust and relationships that constitute the basis of digital communication and the creation of online offers. To Lin (2004), it is possible to categorize potential or real online customers according to their specific needs and characteristics with the use of present information technology. These groups are identified and distinguished thanks to various online methods, and then one or more are selected as the target market of the corporation. A special marketing mix is developed according to the characteristics of the groups in order to meet the needs of the groups determined as the target market and to establish a long-term, positive interaction with them. Some of the online methods used in categorizing current or potential customers are listed below (Simmons, 2007: 549):

- To request the demographic information of customers and how to contact them
- To access the electronic addresses of internet users by corporations
- "Cookie" applications that web preferences are followed and can be accessed by corporations (Prabhaker, 2000: 163)
- Google analystic
- Online surveys
- Databases

With these methods listed above, corporations can analyze customer data collected online whenever they desire, customize customer preferences, and provide effective online customer relationship management.

Marketing Communication

Relationships with customers are at the center of customers' preference for the corporation and customer loyalty. Establishing and maintaining relationships with customers is carried out within the scope of marketing communication notions. Rowley (2004) regards the Internet as a marketing communication channel and mentions that the Internet offers the opportunity of branding by providing the following three significant conditions.

Creating Assets
Creating an asset is the process of introducing itself and its products to the customers, creating an identity and personality. The important thing in the digital environment is to ensure that the customer continues to stay on that screen when s/he encounters the web page of the corporation. The goal of a website is to inform, persuade, and remind customers about the corporation and its products and services.

Building Relationships
Relationships emerge as a result of the cooperation of customers and corporations, and as interaction increases, relationships become stronger and more sustainable. Internet enables growth with customer service interaction, customer feedback forms, user registration and the collection of cookie data, and in-house customer interaction. Corporations can use it to customize and personalize their marketing communications, and customers with the information and data collected. Corporations that analyze customers can also prefer to diversify the quality and scope of their products and services in favor of profitable customers. Customer support and service, which is one of the opportunities created by the Internet to communicate and establish relationships, increases the value of the interaction with the customer, and therefore improves customer satisfaction.

Creating Mutual Value
Mutual value emerges when customers and corporations cooperate that are beneficial to both of them. Corporations, which engage the attention of online communities and manage them, have a unique opportunity to understand and learn from their customers. These online communities can be personal communities, extended communities, or groups of interest.

Corporations aim to create more specific communication messages about products and services that will be beneficial to them and their customers. The goal of these messages is to evaluate opportunities and gain benefits (Simmons, 2007: 550). When a digital corporation has a community and preferably some information about that community, this present community can give its information to other corporations. For example, the advertising fees, which a corporation may charge for placing banner ads on the web page, may depend on the size of the group that the website is expected to attract.

Interaction

It is defined as the opportunity for people and corporations to communicate directly with each other regardless of distance and location. In digital shopping, customers interact through the environment identified with the notion of "machine interaction". The environment where this interaction occurs is the "websites" of corporations. Websites with a high level of interaction enable customers to respond positively. In terms of corporations, a website with a high level of interaction enables the realization of marketing communication at the point of identifying and meeting customer needs. "To attract customers to the website with various links and to make them busy within the site" (Parsons et al., 1998: 34), to collect their inputs, to respond directly to individual customers using the collected data are the processes that should be done to improve the interaction of a website. With a successful interaction application, it is possible to personalize the content and functionality, which are the same for all people, in accordance with individual preferences and experiences.

Content

In digital branding which is an effort to attract online customers with unique messages, functionality, and content (Ibeh et al., 2005: 355), web pages where customers interact with the corporation should be designed as simple as possible. The purpose of a customer when s/he enters the website of the corporation is to gather information about a specific product-subject or to carry out a transaction such as purchasing a product-service. For this reason, the main duty of the corporation is to make the website design as simple as possible to allow the customer to accomplish her purpose (Taylor & England, 2006). Criteria such as design simplicity of websites, graphical user interface designs, and responding times at the point that users can access content without waiting for more than a few seconds affect the digital brand perceptions and behaviors of the customers.

Conclusion

In the new economic system, which is characterized by digitalization and globalization, the share and importance of shopping in the digital environment gradually increases. Branding becomes highly important in the digital shopping environment, which occurs rapidly due to its nature, reduces physical interaction, and product qualities and benefits can be conveyed online (Rowley, 2004: 131). The emergence of internet technology, particularly the effects of real-time interaction and the presence of many corporations in the market have made branding more complex and dynamic. The digital environment has brought along low costs as well as opportunities to support different marketing activities (Bhatt & Emdad, 2011: 78). Accordingly, the success of digital brands depends on creating an extended customer base, personalizing customers, acting early, creating and offering advantages, providing quality products and services at lower costs, serving internationally by considering local requirements, and developing communication language—logistics—infrastructure systems. (Ibeh et al., 2005: 1270). In the context of these factors, the images of the corporations in the eyes of the customers, which can manage a successful branding process in the digital environment, can make them the market leaders of their sectors. It is thought that the image of a corporation, which successfully manages the branding process in the digital environment, is also successful. A successful image to be created in the digital environment is created through websites that interact with customers.

It is known that many factors such as the easy connection of websites, correct configuration of content, security features, and the impressions on customers affect the image of the corporation. It is significant in the process of corporate image formation that the web sites customers interact with the corporations and graphical user interfaces are designed in a way that allows users to reach the content without waiting for more than a few seconds and to allow the customers to accomplish their purposes.

It is mentioned by professionalists that a significant factor that helps to increase the perceived leadership of a brand is the trust in the brand

and the information obtained by customers to build trust (Baek & King, 2011: 262). The desired information can be obtained with daily communication or additional communication tools such as search engines, and this rich information helps to increase the credibility and perceived leadership of a brand (Chang & Ko, 2014: 65). Therefore, data such as which search engines the websites of the corporations serving in the digital environment are included, what the traffic density is, whether the corporations are on social media and at what rate they are followed can be taken into account among the data that supports their leadership.

A corporation with a positive image affects the decision-making process of the customers positively, while the images of corporations with a negative image can affect the purchasing behavior of the customers negatively. Brand loyalty emerges in the customer who meets the face of the corporation in the digital platform and tests the satisfaction level at every phase as a result of mutual interaction. As loyal customer continues to visit the same website, it may be difficult for competitors to enter or hold on to the digital market. This enables not only successful corporations to become market leaders in terms of market share and financial terms, but also affect to maintain their present leadership (Batu & Kutlu, 2020: 56).

References

Aaker, D. A., & Keller, K. L. (1990). Consumer evaluations of brand extensions. *Journal of Marketing, 54*(1), 27–41.

Aaker, A. D. (1991). *Managing brand equity*. The Free Press.

Aaker, D. A. (1996). Measuring brand equity across products and markets. *California Management Review, 38*(3), 102–120.

Alwi, S., & Faridah, S. (2009). Online corporate brand image and consumer loyalty. *International Journal of Business and Society, 10*(2), 1–19.

Aktuğlu, I. (2004). Marka Yönetimi, Güçlü ve Başarılı Markalar için Temel İlkeler. İletişim Yayınları.

Andrei, A.-G., & Talabă, M. E. (2010). Brand leadership İn the contemporary economy, studies and scientific researches. *Economics Edition, 15*, 309–313.

Baek, T. H., & King, K. W. (2011). Exploring the consequences of brand credibility in services. *Journal of Services Marketing, 25*(4), 260–272.

Bat, M. (2012). *Dijital Platformda Sosyal Medyanın Stratejik Kurumsal İletişime Etkis, (Unpublished Doctoral Dissertation)*. University of Ege.

Batu, M., & Kutlu, Ö. (2020). UMTEB international congress on vocational & technical sciences-X, (Edt. Hakkı Çiftçi, Zhanuzak Alimgerey),

Dijital Ortamda Kurumsal Markaların Algılanan İmajlarının Bulundukları Yere Katkısı, Nakhcivan-Azerbaycan.
Bhatt, G. D., & Emdad, A. F. (2001). An analysis of the virtual value chain in electronic commerce. *Logistics Information Management, 14*(1/2), 78–85.
Chang, Y., & Ko, Y. J. (2014). The brand leadership: Scale development and validation. *Journal of Brand Management, 21*(1), 63–80.
De Chernatony, L., & Dall'Olmo Riley, F. (1998). Defining a 'brand': Beyond the literature with experts' interpretations. *Journal of Marketing Management, 14*(5), 418–424.
Conger, J. A., & Kanungo, R. N. (1987). Toward a behavioral theory of charismatic leadership İn organizational settings. *Academy of Management Review, 12*(4), 637 –647.
Ediz, E. (2005). *Kurumların Uluslararası Rekabet Stratejileri ve Uygulamalı Bir Araştırma*, (Unpublished Master's Dissertation) University of Trakya, Edirne.
Fidan, M., & Summak, M. E. (2001). İşletme Yönetim Felsefesi Gelişimini Etkileyen İki Kavram: Marka ve İmaj Yönetimi. *Selçuk İletişim Dergisi, 1*(4), 29–34.
Göl, G. (1995). *Gümrük Birliğine Geçişte Rekabet Gücünün Belirlenmesi ve Rekabetçi Pazarlama Stratejileri, (Unpublished Doctoral Dissertation).* University of Marmara.
Hague, P., & Jackson, P. (1994). *The power of industrial brands*, e-book: McGraw Hill.
Hwang, J.-S., McMillan, S. J., & Guiohk, L. (2003). Corporate web sites as advertising: An analysis of function, audience, and message strategy. *Journal of Interactive Advertising, 3*(2), 10–23.
Ibeh, K., Luo, Y., & Dinnie, K. (2005). E-branding strategies of internet companies: Some preliminary insights from the UK. *Journal of Brand Management, 12*(5), 355–373.
İlgüner, M. (2006). *Türkiye'de Marka Yaratma ve Yaşatmanın Altın Kuralları.* Rota Yayınları.
Kapferer, J. N. (2008). *The new strategic brand management.* Kogan Page.
Keller, K. L. (2003). *Strategic brand management.* Prentice Hall.
Knapp, E. D. (2000). *Marka Aklı,* (A.T. Akartuna, Trans.). Mediacat Kitapları.
Konecnik, M., & Gartner, W. C. (2007). Customer—based brand equity for a destination. *Annals of Tourism Research, 34*(2), 400–421.
Kotler, P. (2000), *Pazarlama Yönetimi.* Beta Yayınevi.
Kuzic, J., Giannatos, G., & Vignjevic, T. (2010). Web design and company image. *Issues in Informing Science and Information Technology, 7*, 99–108.
Lin, C. A. (2004). Influencing the online consumer's behavior: The web experience. *Internet Research, 14*(2), 111–126.
McQuade, S., Kierzkowski, A., Waitman, R., & Zeisser, M. (1996). Marketing to the digital consumer. *The McKinsey Quarterly, 2*(3), 180–183.

Merrilees, B., & Fry, M. L. (2002). Corporate branding: A framework for e-retailers. *Corporate Reputation Review, 5*(2), 213-225.
Miller, Karen, W., Mills, Michael, K. (2012). Contributing clarity by examining brand luxury in the fashion market. *Journal of Business Research, 65*, 1471–1479.
Moilanen, T., & Rainisto, S. (2009). *How to brand nations, cities and destinations. A planning book for place branding, ebook*. Palgrave MacMillan.
Neelotpaul, B. (2010). A study on interactivity and online branding. *Advances in Management, 3*(3), 13–17.
Pappu, R., Quester, P. G., & Cooksey, R. W. (2005). Consumer-based brand equity: Improving the measurement–empirical evidence. *Journal of Product & Brand Management, 14*(3), 143–154.
Parsons, A., Zeisser, M., & Waitman, R. (1998). Organising today for the digital marketing of tomorrow. *Journal of Interactive Marketing, 12*(1), 31–46.
Perry, A., & Wisnom, D. (2003). *Markanın DNA'sı*, (Z. Yılmaz, Trans.). MediaCat Yayınları.
Porter, M. E. (1985). *Competitive advantage: Creating and sustaining superior performance*. The Free Press.
Porter, M. E. (2000). *Rekabet Stratejileri; Sektör ve Rakip Analizi Teknikleri*, (G. Ulubilgen, Trans.). Sistem Yayınları.
Prabhaker, P., & R. (2000). Who owns the dijital consumer? *Journal of Consumer Marketing, 17*(2), 158–171.
Rao, A. R., Qu, L., & Ruekert, R. W. (1999). Signaling unobservable product quality through a brand ally. *Journal of Marketing Research, 36*(2), 258–268.
Rowley, J. (2004). Online branding. *Online Information Review, 28*(2), 131–138.
Silva, R., Da., Vinhas, Alwi, S., & Faridah, S. (2008). Online corporate brand image, satisfaction and loyalty. *Journal of Brand Management, 16*(3), 119–144.
Simmons, G. J. (2007). I-branding: Developing the internet as a branding tool. *Marketing Intelligence & Planning, 25*(6), 544–562.
Stanton, J. W. (1975). *Fundamentals of marketing*. McGraw-Hill Book Company.
Supphellen, M., & Nysveen, H. (2001). Drivers of intention to revisit the websites of well-known companies. *International Journal of Market Research, 43*(3), 1–12.
Syed Alwi, S. F., & Da Silva, R. V. (2007). Online and offline corporate brand images: Do they differ? *Corporate Reputation Review, 10*(4), 217–244.
Taylor, M. J., & England, D. (2006). Internet marketing: Web Site navigational design issues. *Marketing Intelligence & Planning, 24*(1), 77–85.

Virtsonis, N., & March, S. H. (2009). Brand positioning in the b2b online environment: A case from the UK print industry. *Journal of Brand Management, 16*, 556–570.

Yadin, D. (2002). *The international dictionary of marketing: Over 1000 professional terms and techniques*. https://pdftec.com/am/ama-dictionary-of-marketing-terms

CHAPTER 3

Internet Use and Understanding the Tendency for Media Use in the Post-COVID Period

Ceren Yegen and *Olha Harmatiy*

Introduction

Following the COVID-19 pandemic it goes without saying that nothing will ever be the same anywhere in the world. During the pandemic process, dominant rules, such as social distancing, mask-wearing, and cleaning have become routines of social life, and distance learning and working models have been experienced. The economic, political, and technological effects of the pandemic, which are discussed with various

C. Yegen (✉)
Department of Journalism, Mersin University Faculty of Communication, Mersin, Turkey
e-mail: cerenyegen@mersin.edu.tr

O. Harmatiy
Department of Journalism and Mass Communication, Lviv Polytechnic National University, Lviv, Ukraine
e-mail: olha.v.harmatiy@lpnu.ua

S. R. Mondal et al. (eds.), *New Normal in Digital Enterprises*,
https://doi.org/10.1007/978-981-19-8618-5_3

variants (i.e., Alpha, Beta, Gamma, Delta, Delta Plus), tourism and trade priority policies, and which affect social life in many contexts, have undeniably emerged. The pandemic has also changed the way people access information and news. Parallel to the increase in news consumption, while the Internet is necessary and is used more than ever as an information and news gathering tool, individuals have made extensive use of new media technologies to obtain information and news quickly and reliably during the pandemic process. This situation has led to an investigation of the dynamics of Internet use and the practice of accessing information and news in the post-COVID process. In this study, how individuals will obtain information and news after COVID-19 is investigated and how they are affected by the Internet. Ukraine and Turkey were chosen as samples with purposive sampling to obtain comparative data. The practices of individuals in Turkey and Ukraine to access news and information before and after COVID-19 are evaluated. The survey was conducted online. The findings obtained from the survey study are subsequently discussed.

The Impact of a Pandemic and Sustainability

The COVID-19 pandemic has had effects that have been seen and felt on a global scale. It has had profound psychological and social consequences at many levels of the global population (Wallinheimo and Evans, 2021). The impact of digital technologies on the way people conduct their life and work has increased during and after the crisis. The pandemic has had an instantaneous impact on the fiscal performance of digital substructure companies, which are discussed in positive and negative contexts at the point of service. Internet platforms are thought to have performed better in the economic downturn during the pandemic process (Economic Impact of COVID-19 on Digital Infrastructure, 2020). Economies, social structures, and governments have been given a tough test during the pandemic process. COVID-19 is wreaking havoc around the world (Baldwin & Weder di Mauro, 2020). According to the Edelman Trust Barometer, 2020 data, even brands and companies have played a substantial role in the pandemic process, and it is significant to use resources and creativity to make a difference. According to the data, how well a brand responds to the crisis during the pandemic process has a gigantic impact on the probability of people purchasing that brand in the future, and brands and companies that put their earnings before human health in the process risk people losing trust in them. For national policymakers, it is a question of how to realize significant changes in economic policy in response to the pandemic, while also 'rebuilding it better' (Ashton-Hart, 2020). According to Enria et al., (2021) the confidence of the public and the strength of the fight against the epidemic are important in the strategic success of governments.

Situations such as the closure of schools, the cessation of flights, the end of crowded gatherings, and the closure of workplaces during the pandemic have brought social as well as physical problems. There were remote business models before the pandemic, but remote models became a necessity with the pandemic. Many sectors have had to experience 'remote business models'. The International Labor Organization published the 'Distance Work Practice Guidelines in and after the COVID-19 Environment' in July 2020 (ILO, 2020: 1). 'Distance learning' has also been a highly accepted compulsory experience for the continuation of education during the pandemic process. In many countries, education has continued remotely at many levels.

It should also be said that the COVID-19 pandemic has changed the way the global health community accepts and uses digital health technologies. Innovative solutions have been revealed in the fight against COVID-19, with many health systems applying more digital health solutions than before during the COVID-19 process (thinktech STM Technological Thinking Center, 2020). The World Health Organization (WHO) has also engaged in many activities in the context of crisis communication and general public health. The findings of one study show that the information from WHO as part of COVID-19 is found to be quite reliable by the public who act accordingly. According to research, high familiarity and commitment were observed in most countries (Varghese et al., 2021).

With the effect of COVID-19, a transition to a new world order where sustainable technologies are at the forefront is foreseen. It is said that the basis of the changes will be sustainable technologies in this new period, where interest in the circular economy is predicted to increase, to the extent that COVID-19 has necessitated transformation for both the business world and social life and forced the global economy to create a completely new s-curve. The linear economy based on 'buy, make, consume' is causing social and environmental consequences that are increasingly unsustainable. It is thought that, in the future, companies will prefer to take a proactive role in creating long-term value by taking more inclusive and more sustainable steps (EY Building a Better Working World, 2020).

COVID-19 and the Media Industry

The COVID-19 pandemic has affected several social fields, including the media and journalism, with news becoming a valuable resource for individuals since the beginning of the global health crisis. According to Nielsen (2020), the Coronavirus crisis has had a severe impact, not only on health and communities, but also on the news media. According to Nielsen, while most of the research presented in the Digital News Report 2020 was collected before the virus hit many countries, the findings from their work since then show that the crisis has created a long-term structural focus on a more digital, more mobile, and more platform-heavy media (Reuters Institute for the Study of Journalism Digital News Report 2020). Therefore, examining the dynamics of information consumption in this process is extremely important both in terms of its ability to transform the media and its place in a democracy. A study aiming to analyze the impact of COVID-19 on news consumption, citizens' trust in the media and their ability to detect fake news has reached important results. The results of the study, which conducted an analysis based on secondary data from the online surveys of the Pew Research Center's American Trends Panel in the USA, and which compared data before and after the outbreak, confirm the effect of the pandemic on the media. The findings of the study indicate that substantial developments have emerged, such as the revival of the role of the traditional media, especially television, and the reconnection of citizens, who are generally distant from information, with the news. The research reveals that the spread of disinformation has increased during the pandemic process and that the ability of citizens to detect news that seems to be completely fabricated during the epidemic has also increased (Casero-Ripollés, 2020).

According to data evaluated by a study examining the three reports, due to the need to seek information in times of crisis, there have been remarkable increases in the use of both traditional and new media worldwide. Among countries, the most reliable sources have been the WHO website, the Center for Disease Control and Prevention (CDC) or government websites. In the reports examined, an increase is observed in the number of subscriptions to paid subscription platforms when the periods before and after the epidemic are compared (Bozkanat, 2021: 227–228).

Internet Usage During the Pandemic Process

There are divergences in the use of the internet and social media in terms of age, gender, culture, and many others (Gómez-Galán et al., 2020; Olinto et al., 2021). For example, according to a study (Pew Research Center & April, 2021), most Americans use YouTube and Facebook. However, the use of Instagram, Snapchat, and TikTok, which are other popular applications, is especially common among adults under the age of 30. In Turkey, the number of social media users increased by 6.0 million between 2020 and 2021.[1]

In times of crisis, the need for information and news increases, and individuals enter a more intense information flow. During the pandemic process, individuals are in need of more information and news. There has been serious information pollution with the disinformation observed, especially in social networks, and this has also affected trust in the media. The massive increase in Internet traffic due to the COVID-19 pandemic in a number of countries shortly after the outbreak points to the digital momentum caused by the pandemic (OECD, 2020).

At the very beginning of the state of emergency (March 2020), a questionnaire applied to 244 people in Portugal attempted to determine how people access information regarding COVID-19, how they are critical of several sources, and how they evaluate the reliability of different media. Research reveals certain implications, such as strong exposure to incidental and actively informative content, and privileged traditional media as the main source (Ferreira & Borges, 2020). In one study aimed at examining the effect of quarantine on Internet use by adolescents, compared to their pre-pandemic habits, the relationship between gaming addiction, Internet use, and COVID-19 concerns were also investigated. The results show that adolescents have increased their use of social media sites and streaming services in general. Regardless of the country of dwelling, the outbreak seems to have had a pointed impact on adolescents' Internet use and psychosocial well-being (Fernandes et al., 2020).

A report examining how Australians learn about COVID-19, what sources they find credible, and the impact of heavy news coverage on their well-being, reveals that most Australians are concerned regarding the

[1] Digital 2021: Turkey, 11 February 2021, Simon Kemp, https://datareportal.com/reports/digital-2021-turkey.

COVID-19 outbreak and are experiencing an increase in news consumption. According to the report, anxiety drives news consumption, and those who are anxious consume more news than before. In fact, 78% of anxious individuals said that their news consumption had increased during the pandemic. In general, Australians now access more news than usual. The quarantine period has led to an increase in news and media consumption among women and youth. According to the report, more than two-thirds (70%) of Australians said they have accessed news more than once a day since the COVID-19 outbreak. In addition, half of Australians (52%) use television as their main source of news regarding COVID-19 (COVID-19: Australian news and Misinformation, 2020).

According to Ofcom's research, attempting to understand how people receive news and information about the crisis in the context of the pandemic, half of the respondents said they had encountered false or misleading information concerning COVID-19 in the third week of the lockdown. The most common misinformation faced by individuals has been 'theories linking the origins or causes of COVID-19 to 5G technology'. In terms of reliability, official sources and traditional publishers have come to the fore, and the use of fact-checking has increased over time (Ofcom, 2020).

Here are some of the results of a study that focuses on how people in six countries access news and information about COVID-19: News consumption has increased in all countries. In many countries, many people have used social media and search engines to get information and news about the Coronavirus. The low level of formal education differs in trusting and turning to news organizations or social media and messaging applications for news and information about the Coronavirus (Misinformation, Science, and Media, April 2020, Navigating the 'Infodemic').

A recent study, conducted within the scope of Community-Based Migration Programs and carried out by the Turkish Red Crescent together with the IFRC, also provides current data. The findings of the study reveal that community members in all regions have a high level of awareness (96.5%) regarding COVID-19 in many contexts. The most popular source of information concerning COVID-19 among respondents is television (66.4%). Television is followed by government officials (38.7%), Facebook (34.8%), health workers (31.4%), websites (23.8%), and family and friends (43.3%). While the use of Facebook (24.5%) and WhatsApp (11.1%) to access information is slightly higher among immigrants, the tendency to access information from television programs (35.8%) is more common among local communities (Knowledge, Attitude and Practice (Cap) Evaluation Study on COVID-19, Community Based Migration Programs, September 2020).

As a result of a study conducted in Turkey, it was seen that a significant portion of university students had difficulty in controlling their Internet use during the COVID-19 pandemic, and that their Internet use negatively affected their relationships with family members, created problems in their social interactions, and had positive effects on their academic lives (Baltacı et al., 2020). According to the results of the Community Volunteers Foundation-Youth Needs Survey (2020), young people stated that they had psychological difficulties due to the increase in social distancing during the pandemic process, and the fact that virtual meetings took the place of physical meetings.

Using the Internet to Access Information and News: A Comparative Analysis in the COVID-19 Process

In order to support the theoretical provisions of the research with empirical data, as well as to obtain comparable data related to the identified theme, a survey study had been done so as to examine the dynamics of Internet use, and information and news practices of individuals in the Ukraine and Turkey in the pre-COVID and post-COVID period.

We chose the online survey method because, according to scholars' view, it is suitable to collect data systematically from a number of organizations, individuals, or other institutions of interest (Given, 2008: 847). In addition, online surveys are becoming extremely popular as they are really easier for researchers to conduct and for interviewees to take part in. Hence, they are not only very convenient but also cost-effective. Moreover, today it is a good method for conducting research in the face of the threat of the coronavirus disease spread.

To accomplish the set objectives, we conducted a survey in which citizens of Turkey and Ukraine took part. The sample is random, and although it is not representative, it is still a better overview of the presented topic than an individual expert's opinion. The questionnaire for this survey consists of 20 closed and mixed questions. Answering the latter (mixed) ones, respondents could provide their own opinion if they think that the proposed list of answer options does not cover the question content. Additionally, surveyees were able to select multiple answers to some questions.

The questionnaire was aimed at finding out respondents' reasons, features, and frequency of Internet usage, their attitude toward getting online news and information, the general level of information literacy and digital competence, and the level of trust in Internet sources. In addition, four of these questions were concerned with the socio-demographic data of the surveyees (age, gender, education, area of residence). Moreover, all the respondents had the opportunity to offer their own comments on the topic.

In total, 200 people answered the questions—100 respondents from each country. The majority of respondents are women; they predominate both among respondents from Turkey (56% of women vs. 44% of men) and even more from Ukraine (71% and 29%, respectively).

Regarding education, the profile of Turkish respondents is as follows: Bachelor's Degree (41%), Doctoral Degree (19%), vocational secondary education (16%), general secondary education or less (13%), Master's Degree (11%). In turn, Ukrainian respondents can be characterized as follows: Master's Degree (37%), Bachelor's Degree (36%), vocational secondary education (14%), Doctoral Degree (8%), and general secondary education or less (5%). According to these data, the highest share of the surveyed is university graduates.

Most respondents live in cities with a population of over 100,000 people (76% of Turks and 52% of Ukrainians), and the least (with a predominance of Ukrainians)—in rural areas (8% and 4%). Other indicators are the following: 22% and 12% of respondents from Ukraine and Turkey, respectively, live in cities with a population of up to 50,000; 18% and 8%—in cities with a population of 50,000–100,000.

The majority are young people aged 18–25, an almost equal number from both countries (46% Turkish and 42% Ukrainian respondents). The shares of respondents from Turkey and Ukraine differ in other age categories: at the age of 26–35—35% and 15%, at the age of 36–45—14% and 9%, at the age of 46–55—5% and 24%, respectively. But, in the age group, 56+ only respondents from Ukraine are represented with quite a significant share—10% of the total number. Thus, along with the large representation of youth from both countries, the number of Turkish respondents decreases with age, while the percentage of Ukrainians in the older age categories is relatively high.

Google Forms questionnaire link was provided via email and social network services. Participation in the study was voluntary; the purpose and characteristics of this survey were explained to all respondents. It should be noted that the questions were proposed to the respondents in the relevant language, i.e., the language of the state of each citizen. The survey was conducted in August 2021 simultaneously in Turkey and Ukraine.

Findings

According to the obtained data, the most important purpose for the use of the Internet is to receive news and information, which was indicated by 86% of the Ukrainian and 64% of Turkish respondents (See Fig. 3.1). In our opinion, these results can be explained by the following general reasons. Firstly, the Internet nowadays is an affordable, convenient, fast, and relatively inexpensive way to do it. Secondly, the net contains vast amounts of information of various kinds. As a result, during a pandemic, as well as any other social, economic, or political challenges and upheavals with the increasing need for news and information, the Internet strengthens its role as an information source due to its characteristics.

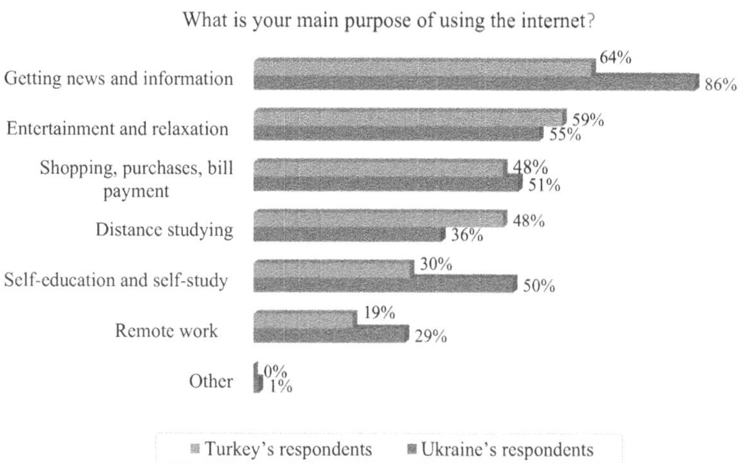

Fig. 3.1 Respondents' main purposes of the internet usage

At the same time, respondents mentioned other purposes for using the Internet, as it was possible for them to offer multiple answers to this question. It is noteworthy that they are almost identical for the representatives of both countries, including entertainment and relaxation; shopping, purchases, and payment of bills; distance learning; self-education, and self-study; remote work.

For receiving news and information, survey respondents typically use social media—80% of Turkish people and 76% of Ukrainians reported this (See Fig. 3.2). The most popular social networks are among respondents from younger age groups; by the results, other socio-demographic characteristics do not show a significant impact. Hence, it can be stated that social networks are a source of information for people of different education, gender, and place of residence, but with a predominance of young people. These results confirm studies that consider social media as an increasingly important source for obtaining information (e.g., Carter & Shields, 2021; Nam & Hwang, 2021; Pearson, 2020).

The survey results prove that Turkish respondents rely mostly on Instagram (72%), Twitter (67%), YouTube (31%), WhatsApp (28%), and Facebook (10%) to receive news and information. For Ukrainians, by

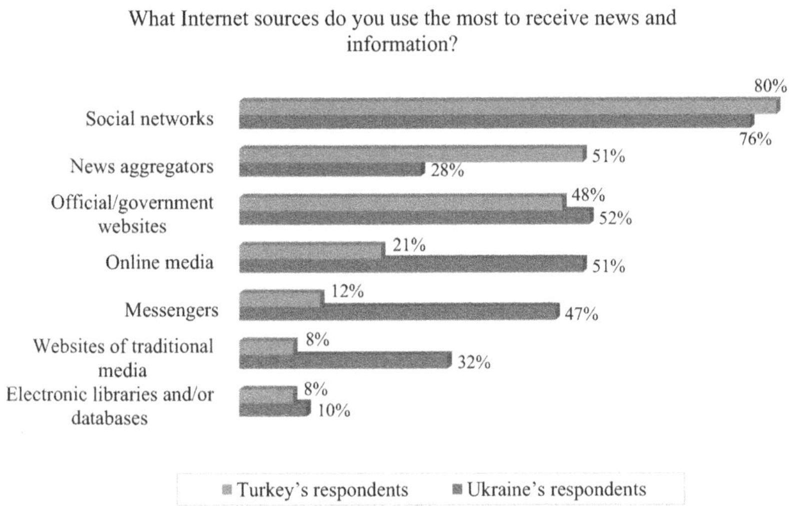

Fig. 3.2 Most popular sources for receiving news and information online

contrast, Facebook tops the list of priority social platforms for receiving news (72%) and YouTube (70%), less popular are Telegram (57%), Instagram (44%), and Viber (30%).

In addition to social platforms, the top 3 Internet sources for both Turkey's and Ukraine's respondents also include official and government websites, as well as news aggregators. Messengers and online media are also quite popular, especially among the Ukrainian audience. Instead, websites of traditional media, electronic libraries, and databases are even less popular.

The importance of government information for citizens during the spread of coronavirus disease has been noted in recent studies. For example, Shaniece B. Bickham and Diane B. Francis, after studying the public opinions of government officials' communication at the time of the pandemic, indicated that people mostly used government websites to obtain the data, and those sites were among the top five informational sources. More than half of those polled pointed out that they confided in the governments at various levels about health care issues (Bickham & Francis, 2021: 196–197).

We received very similar answers to the question about the devices that respondents most often use to access the Internet. Almost all (98% of Turkish people and 93% of Ukrainians) reported that they use smartphones, followed by laptops and PCs with a significant lag. Thus, the results show a significant potential for the development of mobile Internet and network access through mobile devices (smartphones).

Other media and communication scholars draw similar conclusions. For example, Eylem Yanardağoğlu's (Yanardağoğlu, 2021: 149) findings indicate such an overarching common trend as increased mobile news access. In this regard, we can consider both benefits and risks of it. In particular, according to Logan Molyneux (Molyneux, 2018: 634), mobile devices may increase the tendency to prefer snack news as they encourage short and dispersed consumption patterns.

On the other hand, Jakob Ohme (Ohme, 2014: 299) lists the advantages of using a mobile connection from a citizens' perspective. In addition to wide opportunity to get data, it is easily customized to suit the user. Location-based services can provide consumers with valuable information concerning their local area. Besides, the capability of the information service to meet the users' desire to receive only the information they want to result in its higher individualization. Easier access to information on a mobile device with intuitive controls and the possibility to obtain information while wasting some time (especially when getting stuck in traffic jams, waiting in line, etc.) can better explain the trends.

The majority of respondents (53% from Ukraine and 39% from Turkey) are partially satisfied with the quantity and quality of news and information on the Internet (rather 'yes' than 'no'). The share of those who are dissatisfied with it is almost the same: 22% of Ukrainian and 20% of Turkish residents. Therefore, according to the results, the Ukrainian audience is more satisfied with the Internet content.

Additionally, the survey results reveal that Ukrainian respondents trust more in online news and information compared to Turkish (See Figs. 3.3 and 3.4).

As shown in the figures above, the majority of the surveyees trust online news and information partly. In our opinion, that is a paradox: people use the Internet mainly for obtaining news and information though, at the same time, they treat it with considerable distrust. Furthermore, additional related information is collected here: according to the

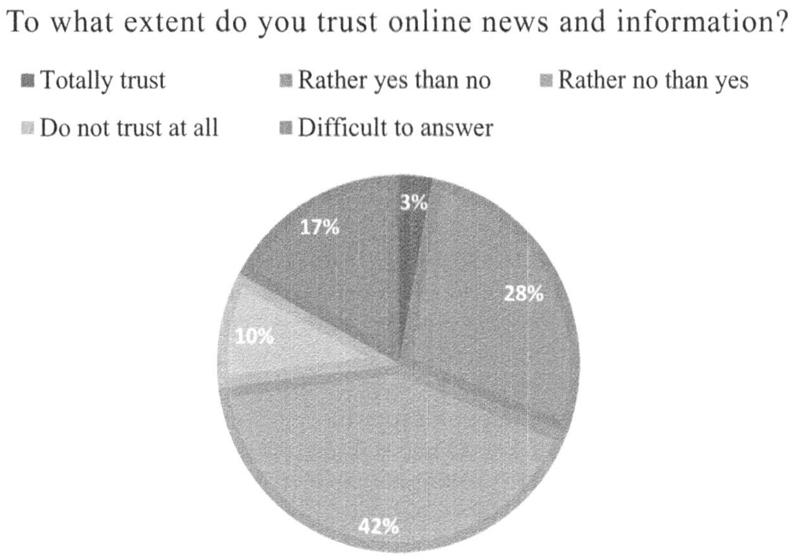

Fig. 3.3 Level of Turkish respondents' confidence in online information and news

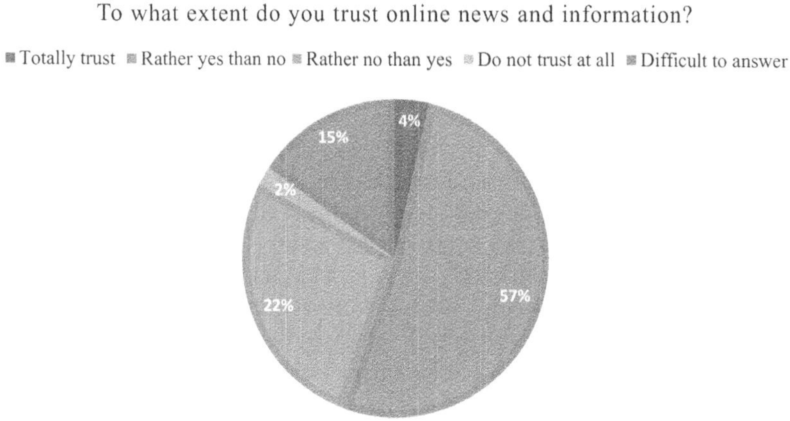

Fig. 3.4 Level of Ukrainian respondents' confidence in online information and news

survey, the question of trust/distrust, unlike other questions, turned out to be difficult, so a large number of respondents failed to answer it.

In this regard, it should be emphasized that a large number of consumers of social media worldwide state that they do not trust social networks either as media sources or as a way to receive news, but they continue using them daily to learn about current news and events (Watson, 2020).

The results also indicate that in case people have doubts about the reliability of information obtained on the Internet, they tend to verify it in other sources—this was noted by 83% of Turkish and 79% of Ukrainian respondents. Other options for Turkish respondents in such situations are to learn more about the expert/speaker (31%), and to refer to fact-checking organizations (26%). Ukrainians have chosen the following options: to take this information as one of the possible versions of events (38%), to think about who benefits from this interpretation of facts and to learn more about the media (both 29%), to learn more about the media (36%). Thus, survey results prove the respondents' critical attitude to the consumption of news and information on the Internet and their attempts to determine the credibility of the messages.

According to the survey results, the desired characteristics of information and news on the Internet are verified facts, interesting content, attractive headlines, photos and videos, presentations of different points of view, and simplicity (See Fig. 3.5).

Regarding thematic priorities, there are common features, and, at the same time, significant differences. In particular, all respondents are interested in recent national and international events (90% of Turkish and 89% of Ukrainian). People are also interested in topics such as medicine and health care—almost half of respondents from both countries (with a larger share of Ukrainians: 55% vs. 44%); science and technologies, on the other hand, are dominated by Turkish citizens: 47% and 32%; politics—44% of Turks and 42% of Ukrainians. Instead, culture and art (30%), sports (30%) ecology (15%), tourism, and travel (14%) are less popular topics for the Turkish audience. In contrast, Ukrainians prefer the topic of culture and arts—half of the respondents, as well as they are highly interested in local

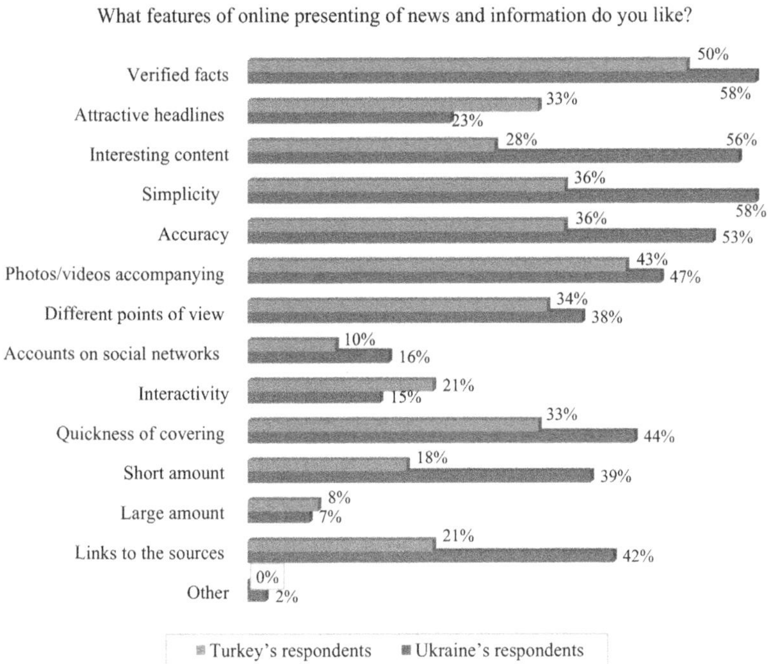

Fig. 3.5 Preferred features of online presenting of news and information

events of their city/town/village (63%); economics (28%), ecology (32%), and sports (25%) are of least interest.

The survey respondents report that the most common way to interact with information sources and media is by sharing news and information on social networks (49% of Ukrainian and 38% of Turkish respondents), as well as commenting and expressing their own views (51% of Ukrainians and 19% of Turks). This result is obvious because of the fact that the social networks mentioned above are the main source of news and information, and they have significant interactive tools and are the key channel for interactive participation on the Internet. According to Melissa W. Graham (Graham, 2014: 361), the fame of social media is revolutionary on a global scale.

It should be noted that the relationship between an interactivity indicator and age shows that young people predominate among those who interact the most. It is clear that interaction is a structural feature of both new media and modern youth (Harmatiy, 2020: 642). Today's audience is willing to be more active than passive, and such interaction helps change the status of the audience. Chris Wells (Wells, 2014: 616) points out that the logic and structure of information systems are undergoing changes.

In this regard, Johan Östman (Östman, 2012: 1004) emphasizes that adolescents reaching adulthood in the 2000s, unlike most other adult populations, do not consider either the Internet or user-generated content to be 'new media'; rather, they have always been included into their daily media use.

Other ways of interactivity are not so popular, namely, responding to audience surveys, polls, participating in contests, special projects, rating/identifying topics for future materials, asking questions, taking part in discussions, adding/offering own content (text, photos, video) on social important events, assisting in improving the materials' quality. However, a third of Turkish respondents (compared to 7% of respondents from Ukraine) said that they did not interact with internet sources. Instead, a significant proportion of people discuss the information with relatives, friends, or colleagues—22% of the Turkish and even more Ukrainians—46%.

The survey data confirm that people began to spend more time receiving news and information online than before the COVID-19 pandemic. The majority of Turkish respondents (30%) stated that before the pandemic they spent an average of 1–2 h a week on the Internet to receive news and information, but this time increased to 3–5 h (23%) after the pandemic.

A similar pattern is observed among Ukrainian respondents. During the pre-COVID-19 period, the share of those spending 1–2 h a week receiving news/information prevailed (41%). After the pandemic, their number decreased, but the share of citizens who spend 3–5 h online increased (from 16 to 26%), and those spending 9–11 h grew significantly (from 5 to 16%). Thus, the results of the survey show that in the post-COVID-19 period, more people use more time to obtain news and information.

Furthermore, the frequency of use of Internet information resources is also increasing. The percentage of Turkish respondents who generally used the Internet several times a day to receive news/information increased significantly—from 66% (before the pandemic) to 87% (after the pandemic), similar indicator is considered among Ukrainians; it increased from 65 to 79%. This indicator increased the most among young people (in terms of age), people with higher education (in terms of education) and women (in terms of gender), but the place of residence did not have a significant impact.

After analyzing these results, it is clear that after the pandemic, there is a growing interest in receiving information and news online. In addition to the general reasons for Internet usage (such as availability, cheapness, convenience, etc.), the influence of quarantine restrictions and isolation led to its active usage during the pandemic. In the post-COVID-19 period, a lot of people continue to be supporters of the Internet as an information source. We agree with H. Pocheptsov's (Pocheptsov, 2021) point of view that the world fully exists in a new dimension that is information, and it cannot be run back. The Internet is an important and perhaps indispensable component of modern information consumption.

Noticeably, it was explored that a great deal of participants (49% of Turkish and 47% of Ukrainian citizens) evaluated their knowledge and skill levels regarding Internet use at a moderate level. The number of respondents who stated that they had a high level of knowledge and skills regarding Internet use was also high—47% of residents from Turkey and 44% from Ukraine (See Figs. 3.6 and 3.7).

An important finding of the survey is that the majority of people (53% of Turkish and 37% of Ukrainian) want to attend training aimed at improving the use of digital tools. This highlights the respondents' interest in digital media literacy and willingness to improve their knowledge and skills in using the Internet. Today, digital competence and media literacy have become key skills for people. The more information society develops, the more digital competence is required. The capability

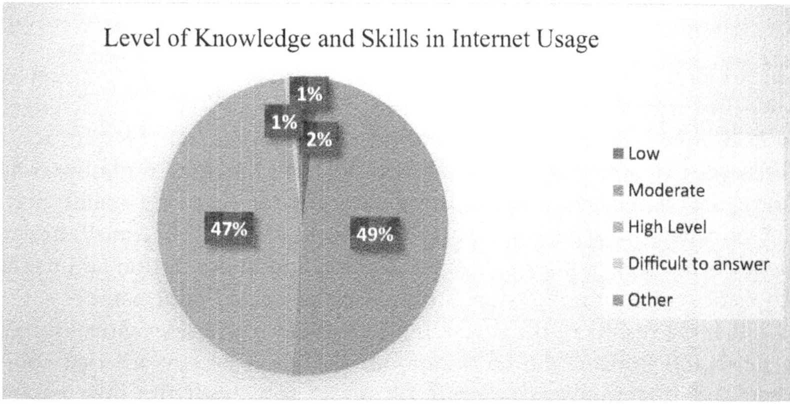

Fig. 3.6 Knowledge and skills of Turkish respondents' internet use

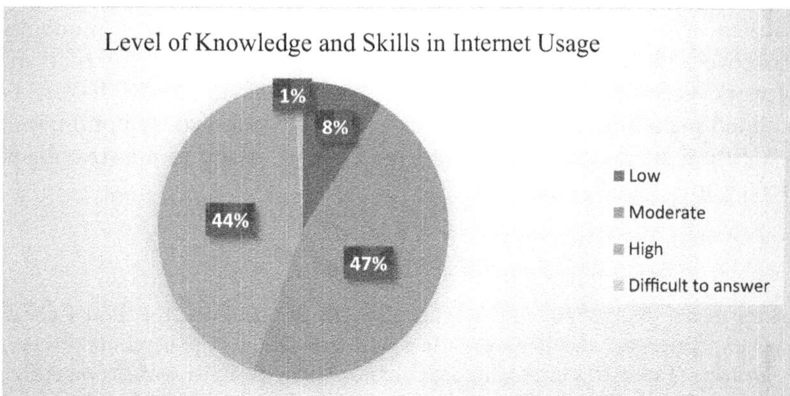

Fig. 3.7 Knowledge and skills of Ukrainian respondents' internet use

to distinguish between true and false news and information, manipulations and trusty communication is becoming a requisite element of digital security (Eger et al., 2020: 85; Ziemba, 2013). Moreover, digital literacy is highly important to be a full member of knowledge society (Martínez-Bravo et al., 2021: 76).

Conclusion

The obtained data and the results of the study that consumption of news and information during the post-COVID period is characterized by common features across the world. As it turned out, despite some differences, people from countries with different histories, traditions, religions, and languages had similar preferences for receiving online news and information. The Internet as the World Wide Web contributes to the globalization of information production and consumption and to the acquisition of common features for getting news and information.

Thus, the results of the research give ground to consider global trends in news and information consumption. It should be stressed that some significant quantitative changes took place, too: both the time people spend to obtain information and the frequency of access to Internet resources have increased. Moreover, the obtained data reflect the qualitative characteristics of information consumption. First of all, social networks, social networks are an important source of news and information, and people, especially the youth, use social media tools frequently. Moreover, the findings confirm the demand for mobile Internet and network access via mobile devices (smartphones). Finally, it can be outlined such important positive trends in information consumption, such as a critical attitude to sources and the citizens' desire to improve media literacy and digital competence.

References

Ashton-Hart, N. (2020). Leveraging Digital Connectivity for Post-COVID Competitiveness and Recovery. Retrieved from https://policycommons.net/artifacts/1159910/leveraging-digital-connectivity-for-post-covid-competitiveness-and-recovery/1718065/on15Dec2022.CID:20.500.12592/k3xrdp.

Baldwin, R. & Weder di Mauro, B. (2020). *Economics in the time of COVID-19.* Edited by Richard Baldwin and Beatrice Weder di Mauro, A CEPR Press VoxEU.org eBook, Centre for Economic Policy Research, ISBN: 978-1-912179-28-2

Baltacı, Ö., Akbulut, Ö. F. & Zafer, R. (2020). COVID-19 Pandemisinde Problemli İnternet Kullanımı: Bir Nitel Araştırma. *Kırşehir Ahi Evran University Journal of Health Sciences Institute* 2020, *1*(3), 126–140.

Bozkanat, E. (2021). Koronavirüs salgınında medya kullanımı: Küresel raporlar üzerinden bir derleme. *Etkileşim, 7*, 216–234. https://doi.org/10.32739/etkilesim.2021.7.125

Bickham, S. B., & Francis, D. B. (2021). The public's perceptions of government officials' communication in the wake of the COVID-19 Pandemic. *Journal of Creative Communications, 16*(2), 190–202. https://doi.org/10.1177/09732586211003856

Carter, J., & Shields, B. (2021). Social media debate position 3: Social media as a credible source of information. *Internet Reference Services Quarterly, 25*(1–2), 49–54. https://doi.org/10.1080/10875301.2021.1940411

Casero-Ripollés, A. (2020). Impact of Covid-19 on the media system. Communicative and democratic consequences of news consumption during the outbreak. *El Profesional De La Información, 29*(2), e290223. https://doi.org/10.3145/epi.2020.mar.23

Community Volunteers Foundation—Youth Needs Survey. (Toplum Gönüllüleri Vakfi - Gençlerin İhtiyaçları Araştırması). (2020). Koronavirüs Pandemi Süreci ve Gençlerin İhtiyaçları Araştırması, https://www.tog.org.tr/wp-content/uploads/2020/07/Koronavir%C3%BCs-Pandemi-S%C3%BCreci-ve-Gen%C3%A7lerin-%C4%B0htiya%C3%A7lar%C4%B1-Ara%C5%9Ft%C4%B1rmas%C4%B1.pdf

COVID-19: Australian news and misinformation by Sora Park, Caroline Fisher, Jee Young Lee and Kieran McGuinness, NEWS & MEDIA RESEARCH CENTRE, This Report was supported by Deputy Vice-Chancellor Research and Innovation Strategic Funds. The project received ethics approval from University of Canberra Human Ethics Committee (No. 2275), Published by the News & Media Research Centre, Canberra, Australia. ISBN: 978-1-74088-500-3.

Economic Impact of COVID-19 on Digital Infrastructure. Report of an Economic Experts Roundtable organized by ITU GSR-20 Discussion Paper July 2020, © ITU 2020 International Telecommunication Union Place des Nations CH-1211 Geneva, Switzerland.

Edelman Trust Barometer 2020. Special Report: Brand Trust and the Coronavirus Pandemic. Timing of Fieldwork: March 23–March 26, 2020, https://www.edelman.com/trust/2020-trust-barometer

Eger, L., Egerová, D., Mičík, M., Varga, E., Czeglédi, C., Tomczyk, L. & Sládkayová M. (2020). Trust building and fake news on social media from the perspective of university students from four Visegrad countries. *Communication Today, 11*(1), 70–88.

Enria, L., Waterlow, N., Rogers, N. T., Brindle, H., Lal, S., Eggo, R. M., et al. (2021). Trust and transparency in times of crisis: Results from an online survey during the first wave (April 2020) of the COVID-19 epidemic in the UK. *PLoS ONE, 16*(2), e0239247. https://doi.org/10.1371/journal.pone.0239247

EY Building a Better Working World. (2020). COVID-19 Sonrası Bilgi Teknolojilerinin Dijital Dünyada Yeni Rolü. 2020 August, https://assets.ey.com/content/dam/ey-sites/ey-com/tr_tr/pdf/2020/08/ey-turkiye-covid-19-sonrasi-bilgi-teknolojilerinin-dijital-dunyada-yeni-rolu--rapor.pdf

Ferreira, G. B. & Borges, S. (2020). Media and Misinformation in Times of COVID-19: How People Informed Themselves in the Days Following the Portuguese Declaration of the State of Emergency. *Journal Media* 2020, 1, 108–121; doi:https://doi.org/10.3390/journalmedia1010008.

Fernandes, B., Nanda Biswas, U., Tan-Mansukhani, R., Vallejo, A. & Essau, C. A. (2020). The impact of COVID-19 lockdown on internet use and escapism in adolescents. *Revista de Psicología Clínica con Niños y Adolescentes* Vol. 7 n°. 3-Septiembre 2020—pp. 59–65. https://doi.org/10.21134/rpcna.2020.mon.2056

Given, L. M. (2008). *Encyclopedia of qualitative research methods*. SAGE Publications.

Gómez-Galán, J., Vergara, D., Ordóñez-Olmedo, E. & Guadalupe Veytia-Bucheli, M. (2020). Time of Use and Patterns of Internet Consumption in University Students: A Comparative Study between Spanish-Speaking Countries. *Sustainability*, *12*, 5087. https://doi.org/10.3390/su12125087.

Graham, M. W. (2014). Government communication in the digital age: Social media's effect on local government public relations. *Public Relations Inquiry*, *3*(3), 361–376. https://doi.org/10.1177/2046147X14545371

Güz, N. & Yanık, H. (2020). Toplumsal Sorumluluk Yaklaşımı ve Güven Bağlamında Medya ve Kamuoyu Araştırmaları Arasındaki İlişki. *Gümüşhane University Faculty of Communication Electronic Journal (e-GİFDER), 8* (1), 298–336. https://doi.org/10.19145/e-gifder.671286

Harmatiy, O. (2020). Media and scientific literacy development within the framework of public engagement with science. *Media Education (mediaobrazovanie), 60*(4), 636–644. https://doi.org/10.13187/me.2020.4.636

International Labor Organization. (ILO). (2020). COVID-19 Ortamında ve Sonrasında Uzaktan Çalışma Uygulama Kılavuzu. Geneva: International Labor Office, July 2020, https://www.ilo.org/wcmsp5/groups/public/---europe/---ro-geneva/---ilo-ankara/documents/publication/wcms_759299.pdf

Knowledge, Attitude and Practice (Cap) Evaluation Study on COVID-19. Community Based Migration Programs, September 2020, Prepared by Turkish Red Crescent Community Center, Staff: Tenadi Gölemerz, Mehmet Akdaş, Semih Paslı, Sayeeda Farhana, Klaudia Jankowska, Turkish Red Crescent Community Based Migration Programs.

Martínez-Bravo, M. C., Sádaba Chalezquer, C., Serrano-Puche, J. (2021). Meta-framework of digital literacy: comparative analysis of 21st-century skills frameworks. *Revista Latina de Comunicación Social*, 79, 76–110. https://doi.org/10.4185/RLCS-2021-1508

Misinformation, Science, and Media, April 2020. Navigating the 'Infodemic': How People in Six Countries Access and Rate News and Information about Coronavirus. Rasmus Kleis Nielsen, Richard Fletcher, Nic Newman, J. Scott Brennen, and Philip N. Howard, The report is based on a survey commissioned by the Reuters Institute for the Study of Journalism and the Misinformation, Science and Media project run with the Oxford Internet Institute and supported by the Oxford Martin School, https://reutersinstitute.politics.ox.ac.uk/sites/default/files/2020-04/Navigating%20the%20Coronavirus%20Infodemic%20FINAL.pdf

Molyneux, L. (2018). Mobile news consumption: A habit of snacking. *Digital Journalism, 6*(5), 634–650. https://doi.org/10.1080/21670811.2017.1334567

Nam, S.-J., & Hwang, H. (2021). Consumers' participation in information-related activities on social media. *PLoS ONE, 16*(4), e0250248. https://doi.org/10.1371/journal.pone.0250248

OECD. (2020). Digital Transformation in the Age of COVID-19: Building Resilience and Bridging Divides. Digital Economy Outlook 2020 Supplement, OECD, Paris, www.oecd.org/digital/digital-economy-outlook-covid.pdf

Ofcom. (2020). COVID-19 news and information: consumption and attitudes. Results from weeks one to three of Ofcom's online survey, Publication date: 21 April 2020.

Ohme, J. (2014). The acceptance of mobile government from a citizens' perspective: Identifying perceived risks and perceived benefits. *Mobile Media and Communication, 2*(3), 298–317. https://doi.org/10.1177/2050157914533696

Olinto, G., Sugahara Pinheiro, S., Bernuci dos Santos, N. (2021). Go, girls! longitudinal data on gender differences in Internet use in Brazil. *IRIE International Review of Information Ethics*, Vol 30. (08/2021), https://informationethics.ca/index.php/irie/article/view/406/422

Östman, J. (2012). Information, expression, participation: How involvement in user- generated content relates to democratic engagement among young people. *New Media and Society, 14*(6), 1004–1021. https://doi.org/10.1177/1461444812438212

Pearson, G. (2021). Sources on social media: Information context collapse and volume of content as predictors of source blindness. *New Media and Society, 23*(5), 1181–1199. https://doi.org/10.1177/1461444820910505

Pew Research Center, April 2021, "Social Media Use in 2021". By Brooke Auxier and Monica Anderson, For Release April 7, 2021.

Pocteptsov, G. (2021, July 5). *The world fully exists in a new dimension—Information, and it cannot be runned back.* Available at https://rezonans.asia/mir-polnocenno-sushestvuyet/

Reuters Institute Digital News Report 2020. Nic Newman with Richard Fletcher, Anne Schulz, Simge Andı, and Rasmus Kleis Nielsen, Reuters Institute for the Study of Journalism / Digital News Report 2020.

Thinktech STM Technological Thinking Center. (2020). COVID-19 Sonrası Sağlık Teknolojisi ve Klinik Karar Destek Sistemleri. Trend Analysis, June 2020.

Varghese, N.E., Sabat, I., Neumann-Böhme, S., Schreyögg, J., Stargardt, T., Torbica, A., et al. (2021). Risk communication during COVID-19: A descriptive study on familiarity with, adherence to and trust in the WHO preventive measures. *PLoS One, 16*(4), e0250872. https://doi.org/10.1371/journal.pone.0250872

Wallinheimo, A.-S.; Evans, S. L. (2021). More Frequent Internet Use during the COVID-19 Pandemic Associates with Enhanced Quality of Life and Lower Depression Scores in Middle-Aged and Older Adults. *Healthcare* 2021, *9*, 393. https://doi.org/10.3390/healthcare9040393

Watson, A. (08.07.2021). Social media as a news source worldwide 2020. Available at https://www.statista.com/statistics/718019/social-media-news-source/

Wells, C. (2014). Two eras of civic information and the evolving relationship between civil society organizations and young citizens. *New Media and Society, 16*(4), 615–636. https://doi.org/10.1177/1461444813487962

Yanardağoğlu, E. (2021). 'Just the way my generation reads the news': News consumption habits of youth in Turkey and the UK. *Global Media and Communication, 17*(2), 149–166. https://doi.org/10.1177/1742766520979729

Ziemba, E. (2013). The Holistic and systems approach to the sustainable information society. *Journal of Computer Information Systems, 54*(1), 106–116.

PART II

Sustainable Information Management and COVID-19

CHAPTER 4

Infodemic as a Sign of Modern Populism: A Discussion on Fake Information About the COVID-19 Pandemic on Twitter

Mikail Batu and Onur Tos

Introduction

Nowadays, with the effect of digitalization, significant developments are observed in the speed of information in the fields of health, culture, politics, and economy. Particularly with the possibilities of web 2.0 technology, people can reach the information they expect on any subject in a short time via computers and mobile devices. However, the accuracy or inaccuracy of this information is frequently not questioned by people,

M. Batu (✉)
Faculty of Communication Faculty, Department of Public Relations and Publicity, Ege University, Bornova, Turkey
e-mail: mikail.batu@ege.edu.tr

O. Tos
Vocational School Marketing and Advertising Department, Tarsus University, Mersin, Turkey

S. R. Mondal et al. (eds.), *New Normal in Digital Enterprises*,
https://doi.org/10.1007/978-981-19-8618-5_4

and the contents on social media networks can be accepted within the framework of the way they are disseminated.

The pandemic has greatly shaken the human race and deeply affected their social life. Discussions have begun regarding the accuracy of the information disseminating in this period. However, the news about the pandemic finds a large number of places in both traditional mass communication and new media tools. The dissemination speed of pandemic information, particularly through new media tools, is at a high level. It is not considered that the contents of the news contain lies, sarcasm, parody, or satire. In this respect, the dissemination of disinformation has increased around the world. This increase causes more false information and content about the COVID-19 pandemic to be shared today and a crisis in the field of health to be evaluated from the perspective of misinformation.

The handling of fake, incorrect, and false information on social media networks about the COVID-19 pandemic is described as infodemic in the literature. Infodemic is defined as the dissemination of false information (misformation) and the inability to access accurate and reliable information in crisis due to the rapid dissemination of fake information. On the other hand, misinformation disseminating on social media networks during the COVID-19 pandemic varies. Many examples can be listed such as the dissemination of a large number of different content about anti-vaccination, the statement that the virus is spread on purpose, the emphasis that it originated from bats and anteaters, and its treatment as a game. In response to the dissemination of misinformation on different and uncontrollable subjects, many institutions, particularly the World Health Organization, work to prevent the dissemination of infodemic on the axis of social media. At the beginning of these studies, different institutions support the rapid dissemination of accurate information. In this context, this study focuses on the shares in social networks, which are considered within the scope of infodemic in the COVID-19 pandemic. The study aims to discuss the contents of the infodemic related to the COVID-19 pandemic and to examine them. In this sense, the confirmatory artificial intelligence application of the www.teyit.org website is applied and the accuracy of the information is discussed.

The Concept of Fake Information as a Sign of Modern Populism

Populism can be considered an ideology that includes political, socio-cultural processes that support democracy in the political processes of countries or can pose a threat (Alkan, 2019, p. 3; Aknur, 2020, p. 420). Populism shows homogeneous features and is shaped by the people. However, it focuses on the meaning of the system between the local people and the elite. In this context, it is shaped on a localist understanding or authority; it can be characterized by the concept of the enemy, which stands next to or symbolizes the opposing side (Mudde, 2004, p. 543; Mudde, 2007, p. 22). Populism can be produced directly or without an intermediary, as well as through institutional-based discourses (Yetkin, 2016, p. 71; Baykan, 2017, p. 162). The recent rise of the concept of populism can be attributed to the economic crises experienced by the countries, as well as to the cultural degeneration and the decline experienced in this direction. Another explanation of populism focuses on regimes and is discussed by examining the methods of democratic practices that are presented (Özer, 2018, p. 41; Uslu, 2021, p. 218). The concept of populism has been on the rise recently and has become a notion that many countries benefit from through discourses.

It is seen that research on populism mostly focuses on public discourse (Yılmaz, 2017, p. 52; Baloğlu, 2019, p. 101). When these studies are examined, it is understood that the discourses differ from country to country. Discourses can be created either personally or popular. It is observed that the activity areas of recent populism have focused on false, distorted news contents and infodemic, particularly in the COVID-19 pandemic. Realities are manipulated and wrong agendas and processes can be created with the disinformation that emerges when the information is manipulated with any intention. Disinformation emerges for any reason in the process of transmitting information/news and causes people to mislead and the truth to be understood differently. For this reason, disinformation also refers to the sum of the processes that need to be fought for the correct transmission and understanding of information (Akyüz et al., 2021a, p. 2818). On the other hand, the way the concept is handled and examined differs according to the various perspectives. For example, fake news content has been identified with concepts such as disinformation, suspicious content, fabrication from a different perspective with the post-truth period (Soysal, 2021). Nowadays, this concept necessitates that it needs to be handled on media platforms by considering the source in communication and the evaluation of its results (Büyükafşar & Özçağlayan, 2019, p. 47). In this context, when the dissemination speed and development in information communication technologies (ICT) are considered, it becomes quite difficult to acquire a skill related to fake news. The ability of individuals to find the right content or to prove the falsehood of fake content depends on their practices in this regard and the clear classification of fake news (Değerli & Değerli, 2018, p. 149). To define fake news, the following items are focused on (Newman et al., 2018):

- Fake news contents that are made to discredit other people or are emphasized to gain more profit,
- News contents that are designed to create a specific agenda and also represent the truth,
- News contents that people cannot find a common ground and have the discomfort of expressing.

The news that is far from reality, to serve specific people, ideologies, ideas, or created by fabrication is defined as fake news. Fake news can be made with the description of an individual, as well as for individuals or institutions to gain extra income and to misrepresent information about themselves (İnce, 2017, p. 553). On the other hand, nowadays, fake news is the depreciation of situations, events, or facts that are believed to be true with the increase in the speed of information dissemination in the face of news that is considered false facts (Kavaklı, 2019, p. 677). Many factors such as the use of news without citing the source, missing news elements, deliberately and willfully sharing false information, and the creation of content through digital theft result in the creation of fake news (Kutlu & Doğan, 2020, p. 86). Fake news is the news that an institution or person deliberately creates with false and misleading content and shares with its target audience (Lazer, et al., 2018).

When the academic literature is examined, the content type of the information, the status, and reliability of the source, the dissemination speed of the content, the optimization of harming people become prominent to explain and understand the generalizations about fake news (Bıçakcı, 2019, p. 66). However, it is seen that there are 7 types of fake

Table 4.1 Types of content in Fake News

Satire and Parody	*Misleading Content*	*Fake Content*	*Fabricated Content*	*Manipulated Content*	*Incorrect Context*	*Faulty Link*
No intent to harm, the potential to cheat	Distorting information to frame a person or topic	Imitation of real sources with fabricated content	Completely false content that is produced to anger and harm	Manipulation of real information or image for deception	Sharing real content with false contextual information	Title, image, or subtitle that do not support the content

news according to the dissemination of the generally accepted content. The types of fake news according to content types are shown as follows (Wardle, 2017) (Table 4.1):

There are many studies on the dissemination, definition, and content of fake news. It is observed that different codes and themes emerge in these studies, and within this framework, different phenomena are presented in the dissemination of fake news. For example, Tandoc et al. (2017) examine thirty-four academic studies on the definition of fake news published between the years 2003 and 2017. In their study, they determine that fake news contents are shaped by the themes of imitation, distortion, propaganda, satire, advertisement, and fabrication. On the other hand, Vosoughi et al. (2018) find out that fake news disseminates rapidly to large masses in the Twitter-focused fake news studies they conduct, these are found more attractive by individuals compared to the content provided with real news, and individuals increase the rate of dissemination by sharing this news quickly. In addition, Allcott and Gentzkow (2017), who conducted studies on motivational sources related to fake news, find out that news producers bring such news to people to support a certain person or institution, and today's digital platforms increase the dissemination speed of this news. Oxford University Reuters Institute for Journalism Studies conducts a study in 2018 and find out that 70% of Turkey uses the internet. In this study, they examine the tendency of individuals to receive news from Facebook, Youtube, Twitter, and WhatsApp, and as a result, they conclude that the rate of exposure to fake news and content in Turkey is 49% (Filibeli, 2018, p. 457).

New media has become more involved in socio-cultural life, challenging traditional patterns and modes of transmission (Şener, 2018, p. 359; Onursoy et al., 2020, p. 489). The interactivity provided by the new media leads people to the digital world, to search for new information on social media networks, and to disseminate fake or false news content rapidly. In this context, the rapid production and dissemination of misinformation on new media lead to an increase in the opportunities for misuse. Consequently, the appearance of fake or false news increases (Sarıoğlu, 2020, p. 386; Akyüz et al., 2021b, 2021b, pp. 219–220). Thus, incomplete or incorrect information force the public to think incompletely and to be exposed to fake news more frequently (Waisbord, 2018, p. 1867). In this respect, the persuasion of the public, which is more frequently confronted with false news, also becomes easier.

The Concept of Infodemic in the COVID-19 Pandemic

The COVID-19 pandemic has brought a series of challenges that have not been encountered in any of the previous pandemics around the world. One of these challenges is related to the place of new media in the lives of people. In the twenty-first century, with the new media, the problem of information overload and information pandemic needs to be examined in detail (Rathore & Farooq, 2020, p. 162; Vaezi & Javanmard, 2020, p. 1). The dominant feature of the information age is the overload and overproduction of information. The COVID-19 pandemic has led to the emergence of a new phenomenon at the point of information overload, and this phenomenon has been described as an infodemic (Sharma et al., 2020, p. 1446). This crisis in the COVID-19 pandemic is associated with the social media infodemic, misinterpretation of statistical information, conspiracy theories, rumors, and the dissemination of misinformation as a propaganda tool (Stephens, 2020, p. 276). As disinformation increases, it becomes difficult to control the infodemic about the prevention, diagnosis, and long-term complications of COVID-19 (Tentolouris, et al., 2021, p. 154). During the COVID-19 pandemic, it becomes a necessity for the World Health Organization (WHO) to combat infodemic processes in addition to the problems related to the pandemic in the health system. For this reason, the WHO has accelerated the flow of information about the pandemic in a controlled manner (Siebenhaar et al., 2020).

The initial studies on the concept of infodemic are carried out together with the field of research called "infodemiology" and emerge to investigate the gap between expert knowledge and public perception within the framework of researching and examining health-related information (Eysenbach, 2002). When the literature is examined, infodemic is associated with different concepts and processes. The most prominent situations are primarily the misinformation about the coronavirus and its assumed effect and followed by describing the richness of the available information about the pandemic and its powerful effects (Simon & Camargo, 2021, p. 5). Infodemic refers to the false news content examined with the pandemic processes with the outbreak of COVID-19 in 2020. According to Erku et al. (2020), infodemic includes the policies carried out in the treatment processes regarding the COVID-19 pandemic, the concerns and the mistakes and lies in the way government institutions present themselves, and it is seen as the danger that it becomes difficult for the public to distinguish fact from fiction.

In the COVID-19 pandemic, there has been an "infodemic" of disinformation related to information around the world on Twitter, Facebook, TikTok, WhatsApp, and other social media tools. It is said that the virus is a Chinese deception initially, and this discourse rapidly disseminates on social media platforms. Another claim is that hair dryers can be used to destroy the virus (Buchanan, 2020, p. 894). However, WHO has warned people about the infodemic against the abundance of information in the COVID-19 pandemic (Gazendam et al., 2020; Olatunji et al., 2020, p. 152). Despite the warnings, it is seen that social media-based misinformation and disinformation continue to affect the mental processes of different audiences within the framework of the COVID-19 pandemic, and this strengthens the concept of disinfodemic (Jamil & Appiah-Adjei, 2020, p. 2). In addition, institutions that actively use social media claim to have control over identifying and banning harmful content. However, the activities carried out to control the pandemic processes and prevent individuals from being exposed to infodemic are inconclusive. This demonstrates that many institutions are not well prepared for the social media-based infodemic during the COVID-19 pandemic. It is essential that the emergency plans of the institutions provide the functionality to respond to the pandemic as soon as possible (Mourad et al., 2020, p. 2146).

Although lessons can be learned from previous experiences with infectious diseases, the infodemic, which develops in the direction of knowledge, has the feature of changing future behaviors and thinking structures of people (Williams et al., 2022, p. 216). Disinformation has led to the growth of the infodemic and the emergence of anti-vaccination in many countries. As a result of infodemic posts, it is known that actions are taken due to anti-vaccination, especially in Europe and the USA. In this context, it is observed that the anti-vaccination and hesitancy stem from infodemic information about COVID-19 vaccines. The increase in the dissemination rate of this information is due to the lack of public health information and the problem of trust among public officials (Carenas, 2021, p. 1). According to the study conducted by Cheng et al., (2020, p. 8), people apply certain strategies to deal with infodemic situations during the pandemic process and to obtain accurate information. These strategies include collecting adequate information by questioning the information. This questioning is good for the mental health of individuals and positively affects their sleep patterns.

The fight against infodemic in the COVID-19 pandemic varies. Campaigns are made for the COVID-19 pandemic and infodemic situations. In these campaigns, special names explaining the essence of the processes and strong emphasis are placed on the slogans for the dissemination of accurate information (Hu et al., 2020, p. 2). However, the governments of some countries have created decrees to counter the infodemic information pandemic COVID-19 and have tried to take measures through emergency legislation (Radu, 2020). In many countries, to solve the problems related to the concept of infodemic during the COVID-19 pandemic, decisions have been taken to increase digital literacy training that meet the needs of individuals. It is believed that with this training, critical awareness of the pandemic is created and the fight against the infodemic is achieved (Fernández-Torres et al., 2021, p. 1782).

Information Disseminating on Social Networks Related to the COVID-19 Pandemic

The table regarding the accuracy of the information about COVID-19 during May–June (2021) on the website of teyit.org, in which the accuracy of the information/news in Turkey is confirmed, is given below. The date range of May and June 2021 is important for the COVID-19 pandemic in Turkey and in the world. This is because summer implementations have begun in this period and restrictions have been removed. In addition, due to the change of the season, the idea that the pandemic is over and there will be no more restrictions has spread exceedingly.

Claim	*Confirmation Status*	*Confirmation Result*
Data for May 2021		
COVID-19 vaccines contain harmful nanoparamagnetic particles that cause magnetization	X	There is no evidence for the claim, and the referred research is about DNA vaccines. There is no DNA vaccine among COVID-19 vaccines
Documents obtained by intelligence do not conclude that the virus results in the laboratory	+	The document shows lab workers, who applied to the hospital in November 2019, may have had COVID-19. There are multiple possibilities in this respect, but all of them do not point out a leak
The COVID-19 vaccine causes heart inflammation in some young people	X	The CDC has not declared that the vaccine causes inflammation. Investigations have continued. Heart inflammation cases are not above normal
Scientists also have a major role in sharing the most accurate information about COVID-19	+	In order not to fall into the trap of dangerous wrong health measures, it is essential to convey the right information with the right attitude
It is claimed that cases and death rates have increased by 100% in countries where COVID-19 vaccines are made	X	Vaccines do not claim to be 100% effective, and a significant reduction in cases has been observed in countries with high vaccination rates
There are no data to show that the vaccine harms infants in nursing mothers and pregnant women	X	Studies are conducted by pharmaceutical and vaccine companies for safe and effective vaccine implementation

(continued)

(continued)

Claim	*Confirmation Status*	*Confirmation Result*
Social distancing does not work	X	Researches indicate that the use of masks in closed areas and the need to ventilate the indoor area frequently are significant, as social distance merely is not enough
Children are forced to be used as vaccine subjects	X	Studies are conducted with the permission of parents or legal guardians
Managing the pandemic process with transparency is of great importance to cope with COVID-19	+	Among the 12 countries that are examined, free and prosperous countries perform better in dealing with the pandemic
It is claimed that the use of masks causes inflammation or cancer in the lungs	X	There is no scientific evidence to support mask-related lung inflammation or cancer
The British Science Board has not declared that a policy of fear is pursued over COVID-19	X	One of the teams that give recommendations to the scientific board has declared that in March 2020, they recommended increasing the perception of danger because the precautions were not followed
There is no evidence of nanoparamagnets causing magnetic reactions in COVID-19 vaccines	X	The article presented as evidence for this claim is not about COVID-19 vaccines
Prof. Dr. Mehmet Ceyhan's studies on vaccines are conducted with volunteer participants, just like all vaccine studies in the world	X	Permission is obtained from the parents or legal guardians of the participants
It is claimed that the EU has decided not to issue Schengen visas to those who have the Sinovac vaccine	X	The vaccination certificate application does not include Turkey as they are not an EU member
It has been claimed that it is dangerous to vaccinate children for COVID-19	X	It has been understood that a report, which is later discovered to be false and removed, causes the allegations

(continued)

(continued)

Claim	*Confirmation Status*	*Confirmation Result*
Italy plans to revive tourism by clearing some islands from the coronavirus	+	With the vaccination of citizens living on these islands, Procida has become Italy's first coronavirus-free island. Some politicians say it puts other people's lives at risk
In the video, President Erdoğan announced that there will be a restriction on May 19	X	The video shared the claim that President Erdogan has announced that there will be a restriction on May 19, 2021, belongs to the year 2020

Claim	*Confirmation Status*	*Confirmation Result*
Data for June 2021		
It is claimed that London Municipality has initiated a tender for future vaccine deaths	X	The tender does not indicate that the deaths have been caused by the vaccine or that vaccine-related death has been predicted
It is claimed that the Sinovac vaccine attacks the immune system and paralyzes it	X	The vaccine mentioned in the claim is Sinopharm and this vaccine is not applied in Turkey. There is no known link between Guillain-Barré syndrome and COVID-19 vaccines
It is claimed that extra supplements should be taken to reduce the side effects of COVID-19 mRNA vaccines	X	Most vaccine side effects go away within a few days. If it does not pass, you need to go to the hospital
President Erdoğan: "COVID-19 vaccines are applied in Britain for 100 pounds"	X	The vaccine is free for every citizen in Britain
In Turkey, the current consent form shows that the phase studies of the COVID-19 vaccine have not been completed	X	The form in the claims is outdated. The third phase studies of the COVID-19 vaccines applied in Turkey have been completed
Football player Christian Eriksen has a heart attack due to the COVID-19 vaccine	X	Eriksen has not been vaccinated at the time of the incident

(continued)

(continued)

Claim	*Confirmation Status*	*Confirmation Result*
The Sinovac vaccine attacks the immune system and paralyzes it	X	The claim is about the Sinopharm vaccine, which is not applied in Turkey and is outdated. There is no statistical causality between COVID-19 vaccines and Guillain–Barre syndrome
Those who are vaccinated against COVID-19 are not taken to MRI	X	Vaccines do not contain enough metal to prevent MRI, and having a vaccine does not prevent MRI
London Municipality has initiated a tender for future vaccine deaths	X	The tender does not indicate that the deaths have been caused by the vaccine or vaccine-related death has been predicted. No loss in the world has been determined to be caused by vaccines
Media professionals play a vital role in providing accurate information to the public about vaccine developments	+	The vaccine declaration has been prepared to inform the public about vaccines correctly and to combat the problem of vaccine indecision in the media
There is no evidence to show that spike protein produced by vaccines has a toxic effect on the human body	+	It is important to create an immune response against spike protein in the fight against COVID-19. mRNA vaccines also target this protein
Uğur Şahin's statement that he does not know what the vaccine does, and his images circulate on social media	X	In the image taken from an old interview, Şahin says that he thinks the vaccine will also prevent transmission and spread. Recent studies also confirm this
The claim, that the third phase studies of COVID-19 vaccines are not completed, has created the agenda	X	The third phase studies are completed
The claim, that Turkey is the first in the world in COVID-19 vaccination rate, has created wide repercussions on social media	X	Comparing the population or countries with unequal access to vaccines does not provide accurate data on vaccination rates No central and transparent authority monitors vaccination all over the world

(continued)

(continued)

Claim	*Confirmation Status*	*Confirmation Result*
The claim is that the serial number appears on the bluetooth screen of those who have had the COVID-19 vaccine	X	There is no substance in the content of COVID-19 vaccines that may be related to bluetooth
There is a claim that those who have received the COVID-19 vaccine are not allowed to travel by plane	X	No airline business has such a procedure
The claim, which D. Fauci's leaked e-mail shows that the coronavirus is a biological weapon, has created the agenda	X	The e-mails have been published within the law, the research in the alleged e-mail is about another virus and does not point out a biological weapon
It has been on social media that those who have the COVID-19 vaccine will not be allowed on flights, as the risk of blood clotting increases at high altitudes	X	Those who are vaccinated are taken on flights, and it has not been scientifically proven that COVID-19 vaccines cause blood clots
After a year of the pandemic, the claims that the coronavirus has come out of the laboratory are still on the agenda	+	Factors such as China's attitude against research and the politicization of the issue have created an obstacle to the determination of the origin of the pandemic
The claim, which Montagnier said that those who receive the COVID-19 vaccine will die after two years, has created the agenda on social media	X	There is no evidence that those who develop antibodies will get sick. So far, no case of antibody-related disease has been recorded in those who have been vaccinated

Source (teyit.org, 2022)

Discussion and Conclusion

The COVID-19 pandemic has increased the dissemination of misinformation around the world. The fake information content in the COVID-19 pandemic processes is begun to be evaluated specifically for the infodemic and is aimed to be explained with this concept. The rapid transmission of social media during the pandemic brings about fake information content to reach different audiences in a short time and people experience information pollution about the process. The information pollution in this process directly affects the routine information acquisition and lives of people. The speed in the dissemination of fake information during the pandemic also brings about social media formations on many issues such as the occurrence of unexpected incidents, anti-vaccination, and lynching attempts against healthcare professionals. In this context, many studies have begun on the concept of infodemic.

In their studies focusing on the problem of false content and information from the perspective of health communication, Ünal and Taylan (2017) demonstrate that fake content and false information disseminated in social media environments, especially on health communication, are published with themes such as attracting attention, being popular, increasing click-through rates, and demonstrating an ideological perspective. Bozkurt (2021), in his study, deals with fake and false news contents in the context of infodemic in Turkey during the COVID-19 pandemic and determines that Turkey is in the first place in disseminating false content. In addition, another result of the study is that the content of the news is as important as the genres they have (lie, parody, etc.) and when people are exposed to news content, it is possible to have different consequences that may even result in death. On the other hand, Akyüz (2020) discusses the COVID-19 pandemic through the dissemination of misinformation, and he concludes that the dissemination of misinformation in Turkey increases every day rapidly, and this increase is influenced not only by individuals but also by institutions, politicians, and famous people who use social media effectively. Moreover, he determines that people are exposed not only to false information about the country they live in but also to global-scale fabricated written content, fictionalized images, distortions, and false information through social media. In their study, which handles the production of vaccines and fake news during the COVID-19, Kesgin and Ünlü (2021) indicate that people benefit more from the information shared on social media about vaccines and follow news websites, news about the vaccine on television, and the statements of the Ministry of Health. Furthermore, they conclude that people remember the names of the vaccines in the information they obtained through social media, even if they are wrong, and they find the news from the media reliable but limited in providing information. Gölbaşı and Metinaş (2020) discuss the COVID-19 pandemic and the concept of infodemic in this process. In their studies, they conclude that risk communication should be established and the struggle should be achieved through social participation during the pandemic, the establishment of a flexible communication system over social media and acting with a solid system understanding give good results, all kinds of false information should be defined and value structures should be used in this definition. In this study, it has been seen that the accuracy of most of the news shared on Twitter about COVID-19 cannot be confirmed. In this respect, it can be said that there is great information pollution in social networks. The effect of this information on the audiences can be considered the subject of another study. In the COVID-19 pandemic, in which the concept of infodemic is effective, digital literacy education needs to be developed for the speed of spreading the right information. In this context, suggestions regarding the concept of infodemic, which are discussed within the scope of the study, are listed as follows:

- Suspicious links should be avoided and links that are not believed to be correct should not be clicked on,
- The source of the news should be investigated,
- It should be ensured that the information is confirmed from different centers (local–foreign),
- Information from official institutions such as the Ministry should be followed,
- People should be informed about the concept of infodemic,
- Preparations should be made for the implementation of proactive processes.

In addition to the above, it is not adequate to consider global pandemics such as COVID-19 from a single perspective. In this sense, it is essential to focus on the sustainability of the social order by establishing teams that include different disciplines.

References

Aknur, M. (2020). Almanya'da radikal sağın yükselişi: Almanya için alternatif partisi, popülizm ve demokrasi. *International Journal of Social Inquiry, 13*(2), 415–446.

Akyüz, S. S. (2020). Yanlış bilgi salgını: COVID-19 salgını döneminde Türkiye'de dolaşıma giren sahte haberler. *Akdeniz İletişim Dergisi*, (34), 422–444.

Akyüz, S. S., Gülnar, B., & Kazaz, M. (2021a). Yeni medyada haber güvenilirliği sorunu: Üniversite öğrencilerinin sahte/yalan haberlere yönelik doğrulama refleksleri. *OPUS—Uluslararası Toplum Araştırmaları Dergisi, 17*(36), 2816–2840.
Akyüz, S. S., Kazaz, M., & Gülnar, B. (2021b). İletişim fakültesi öğrencilerinin sahte/yalan haberlerle ilgili görüşlerine yönelik betimleyici bir çalışma. *Selçuk İletişim*, 216–239.
Alkan, M. N. (2019). Hoş geldin kültüründen popülizme. *Savunma Bilimleri Dergisi, 18*(2), 1–22.
Allcott, H., & Gentzkow, M. (2017). Social media and fake news in the 2016 election. *Journal of Economic Perspectives, 31*(2), 211–236.
Baloğlu, U. (2019). Demokrasi, popülizm ve yok olan doğa. *Moment Dergi - Hacettepe Üniversitesi İletişim Fakültesi Kültürel Çalışmalar Dergisi, 6*(1), 99–129.
Baykan, T. S. (2017). Halkçılık ve popülizm: Türkiye vakası ve bir kavramın kullanımları. *Mülkiye Dergisi, 41*(1), 157–194.
Bıçakcı, B. (2019). Post-truth çağında halkla ilişkilerin "hakikat yöneticiliği" rolü: Gıda ve beslenme alanındaki yalan haberlere yönelik stratejiler. *e-Kurgu—Anadolu Üniversitesi İletişim Bilimleri Fakültesi Uluslararası Hakemli Dergisi, 27*(4), 61–78.
Bozkurt, F. (2021). COVID-19 pandemi sürecindeki sahte ve yalan haberlerin bir getirisi: İnfodeminin Türkiye bağlamında incelenmesi. *Uluslararası Halkbilimi Araştırmaları Dergisi - UHAD, 4*(7), 135–151.
Buchanan, M. (2020). Managing the infodemic. *Nature Physics, 16*, 894.
Büyükafşar, M., & Özçağlayan, M. (2019). Haber ve doğruluk ilişkisi: Doğruluk sorununun geçmişi ve haberin dönüşümü. *Düşünce ve Toplum Sosyal Bilimler Dergisi*, (1), 29–49.
Carenas, N. C. (2021). 'Europe and United States vaccine hesitancy': leveraging strategic policy for 'Infodemic' on COVID-19 vaccines. *Journal of Public Health*, 1–2.
Cheng, C., Ebrahimi, O. V., & Lau, Y.-C. (2020). Maladaptive coping with the infodemic and sleep disturbance in the COVID-19 pandemic. *Journal of Sleep Research*, 1–10.
Değerli, A., & Değerli, B. (2018). Medya okuryazarlığı bağlamında sağlık alanındaki yalan içeriklerin sosyal ağlarda yayılımına bir bakış. *4. Sağlık İletişimi Sempozyumu*, (s. 147–168). Antalya.
Erku, D. A., Belachew, S. A., Abrha, S., Sinnollareddy, M., Thomas, J., Steadman, K. J., & Tesfaye, W. H. (2020). When fear and misinformation go viral: Pharmacists' role in deterring medication misinformation during the 'infodemic' surrounding COVID-19. *Research in Social & Administrative Pharmacy, 17*(1), 1954–1963.

Eysenbach, G. (2002). Infodemiology: The epidemiology of (mis) information. *The American Journal of Medicine, 113*(9), 763–765.

Fernández-Torres, M. J., Almansa-Martínez, A., & Chamizo-Sánchez, R. (2021). Infodemic and fake news in Spain during the COVID-19 pandemic. *International Journal of Environmental Research and Public Health, 18*, 1781–1794.

Filibeli, T. E. (2018). Kullanıcı türevli içerik ve yurttaş medyası. *Akdeniz İletişim Dergisi*, (29), 453–469.

Gazendam, A., Ekhtiari, S., Wong, E., Madden, K., Naji, L., Phillips, M., ... Bhandari, M. (2020). The "infodemic" of journal publication associated with the novel coronavirus disease. *The Journal of Bone and Joint Surgery, Incorporate, 102*(13). https://doi.org/10.2106/JBJS.20.00610

Gölbaşı, S. D., & Metintaş, S. (2020). COVID-19 pandemisi ve infodemi. *ESTÜDAM Halkla Sağlığı Dergisi, 5*(COVID-19 Özel Sayısı), 126–137.

Hu, Z., Yang, Z., Li, Q., & Zhang, A. (2020). The COVID-19 infodemic: Infodemiology study analyzing stigmatizing search terms. *Journal of Medical Internet Research, 22*(11), 1–16.

İnce, M. (2017). Türk spor basınında asparagas haber ve spor ekonomisine etkileri. *Karabük Üniversitesi Sosyal Bilimler Enstitüsü Dergisi, 7*(2), 547–563.

Jamil, S., & Appiah-Adjei, G. (2020). Battling with infodemic and disinfodemic: The quandary of journalists to report on COVID-19 pandemic in Pakistan. *Media Asia*, 1–22.

Kavaklı, N. (2019). Yalan haberle mücadele ve internet teyit/doğrulama platformları. *Erciyes İletişim Dergisi, 6*(1), 663–682.

Kesgin, Y., & Ünlü, D. G. (2021). COVID-19 aşısı ve yalan haber: Aşılanma öncesinde bireylerin yalan haber içeriklerini fark etme, takip etme ve teyit etme eğilimlerinin belirlenmesi. *Galatasaray Üniversitesi İleti-ş-im Dergisi*, (35), 32–55.

Kutlu, A., & Doğan, E. (2020). Kesin bilgi, yayalım: Hakikat sonrası çağda yalan haberlere ilişkin y kuşağının tutum ve davranışları. *Akdeniz İletişim Dergisi*, (34), 83–101.

Lazer, D. M., Baum, M. A., Benkler, Y., Berinsky, A. J., Greenhill, K. M., Menczer, F., ... Zıttrain, J. L. (2018). The science of fake news. *Science, 359*(6380), 1094–1096.

Mourad, A., Srour, A., Harmanani, H., Jenainati, C., & Arafeh, M. (2020). Critical impact of social networks infodemic on defeating coronavirus COVID-19 pandemic: Twitter-based study and research directions. *IEEE Transactions on Network and Service Management, 17*(4), 2145–2155.

Mudde, C. (2004). The populist zeitgeist. *Government and Opposition, 39*(4), 541–563.

Mudde, C. (2007). *Populist radical right parties in Europe*. Cambridge University Press.

Newman, N., Fletcher, R., Kalogeropoulos, A., Levy, D. A., & Kleis, R. (2018). *Reuters institute digital news report 2018*. Reuters Institute For The Study Of Journalism.

Olatunji, O. S., Ayandele, O., Ashirudeen, D., & Olaniru, O. S. (2020). "Infodemic" in a pandemic: COVID-19 conspiracy theories in an African country. *Social Health and Behavior*, (3), 152–157.

Onursoy, S., Turan, E. A., Yeşilyurt, S., & Astam, F. K. (2020). Yalan habere karşı tutum ve davranışlar: Üç üniversite örneğinde durum araştırması. *Türk Kütüphaneciliği, 34*(3), 485–508.

Özer, S. (2018). Avrupa'da yükselen popülizm ve radikal sağ: Avusturya örneği ilişkileri. *EURO Politika*, 39–53.

Radu, R. (2020). Fighting the 'infodemic': Legal responses to COVID-19 disinformation. *Social Media + Society*, 1–4.

Rathore, F. A., & Farooq, F. (2020). Information overload and infodemic in the COVID-19 pandemic. *Journal of the Pakistan Medical Association, 70*(5), 162–165.

Sarıoğlu, E. B. (2020). Yalan haber, "post-truth" kavramı ve medya üçlemesi: Geçmişten günümüze gündem belirleyen örnekler. *İnsan Ve Toplum Bilimleri Araştırmaları Dergisi, 9*(1), 377–397.

Sharma, D. C., Pathak, A., Chaurasia, R. N., Joshi, D., Singh, R. K., & Mishra, V. N. (2020). Fighting infodemic: Need for robust health journalism in India. *Diabetes and Metabolic Syndrome: Clinical Research & Reviews, 14*, 1445–1447.

Siebenhaar, K. U., Köther, A. K., & Alpers, G. W. (2020). Dealing with the COVID-19 infodemic: Distress by ınformation, ınfirmation avoidance, and compliance with preventive measures. *Frontiers in Psychology, 11*, 1–11.

Simon, F. M., & Camargo, C. Q. (2021). Autopsy of a metaphor: The origins, use and blind spots of the 'infodemic'. *New Media and Society*, 1–22. https://doi.org/10.1177/14614448211031908

Soysal, M. (2021). Hakikat ötesi, yalan haber ve kolektif belleği çarpıtma dinamikleri bağlamında 6–7 eylül 1955 olayları üzerine bir değerlendirme. *Belgi*, (22), 481–507.

Stephens, M. (2020). A geospatial infodemic: Mapping twitter conspiracy theories of COVID-19. *Dialogues in Human Geography, 10*(2), 276–281.

Şener, N. K. (2018). "Doğruluk kontrol merkezi" ve "yalan haber" kavramlarına ilişkin çeriklerin medyada yansımasının araştırılması. *Akdeniz İletişim Dergisi*(30), 355–373.

Tandoc, E., Lim, Z. W., & Ling, R. (2017). Defining "fake news": A typology of scholarly definitions. *Digital Journalism, 6*(3), 1–17.

Tentolouris, A., Ntanasis-Stathopoulos, I., Vlachakis, P. K., Tsilimigras, D. I., Gavriatopoulou, M., & Dimopoulos, M. A. (2021). COVID-19: Time to fatten the infodemic curve. *Clinical and Experimental Medicine, 21*, 161–165.

Uslu, C. (2021). Bir ideoloji olarak popülizm. *Anemon Muş Alparslan Üniversitesi Sosyal Bilimler Dergisi - Anemon, 9*(1), 217–230.

Ünal, R., & Taylan, A. (2017). Sağlık iletişiminde yalan haber-yanlış enformasyon sorunu ve doğrulama platformları. *Atatürk İletişim Dergisi*(14), 81–100.

Vaezi, A., & Javanmard, S. H. (2020). Infodemic and risk communication in the era of CoV-19. *Advanced Biomedical Research*, 1–2.

Vosoughi, S., Roy, D., & Aral, S. (2018). The spread of true and false news online. *Science, 359*(6380), 1146–1151.

Waisbord, S. (2018). Truth is what happens to news. *Journalism Studies, 19*(13), 1866–1878.

Wardle, C. (2017). *Fake news. It's complicated.* First Draft. https://firstdraftnews.org/articles/fake-news-complicated/ adresinden alındı

Williams, N. L., Wassler, P., & Ferdinand, N. (2022). Tourism and the COVID-(Mis)infodemic. *Journal of Travel Research, 61*(1), 214–218.

Yetkin, B. (2016). Popülizmin gölgesinde siyaeti ve siyasal iletişimi anlamak. *Erciyes İletişim Dergisi - Akademia, 4*(3), 68–82.

Yılmaz, Z. (2017). Popülizm, halk ve demokrasi: Temsili demokrasinin Açmazları ve radikal demokratik bir popülizmin imkânları. *Mülkiye Dergisi, 41*(1), 33–65.

CHAPTER 5

Sustainability, Digital Security, and Cyber Hygiene During the Covid-19 Pandemic

Ceren Yegen, Ali Murat Kirik, and Ahmet Çetinkaya

Introduction

The Coronavirus, which has turned into a global pandemic, was the phenomenon with the most impact in 2020. It has led to the restriction of almost all social spaces and is the subject of much discussion today. As a result of the virus that the whole world still struggles with, daily routines have been changed and transformed. The decreasing course of

C. Yegen (✉)
Department of Journalism, Mersin University, Mersin, Turkey
e-mail: cerenyegen@mersin.edu.tr

A. M. Kirik
Department of Radio, Television and Cinema, Marmara University, Istanbul, Turkey
e-mail: murat.kirik@marmara.edu.tr

A. Çetinkaya
Department of Journalism, Marmara University, Istanbul, Turkey
e-mail: ahmet@marmara.edu.tr

S. R. Mondal et al. (eds.), *New Normal in Digital Enterprises*,
https://doi.org/10.1007/978-981-19-8618-5_5

in-person communication is based on the argument of transformations in communication technologies. Today, most meetings, appointments, and organizations are being held through communication technologies, due to the risk of infection. All individuals face a common dilemma as the capitalist mainstream world order where borders disappeared has been significantly disrupted.

In this process, digital security and cyber precautions have also gained importance. Considering that many things are done through the internet and information and communication technologies, users should be careful about cyber problems. It is quite possible to stand against all these with cyber hygiene. As a set of measures not only for public institutions and organizations but also for individual users, cyber hygiene can provide protection from many malicious purposes during the pandemic process. Many sustainable internet-based practices and actions have unfortunately brought to mind the security vulnerabilities, data security, and individual privacy rights during the pandemic process and opened them up for discussion. For this reason, the concept of cyber hygiene discussed in this study will always be important not only during the pandemic but also in the digital culture and digital world in which the individual lives.

The Impact of the COVID-19 Pandemic

Although the initial information about COVID-19 was limited, several societies implemented strict measures, such as full closure, quarantines to contain the virus, and reduced impact. These applications had varying success with profound socio-economic side effects (Van Damme et al., 2020). COVID-19 has affected numerous professions. For example, it was observed that dentists were in the high-risk group due to an inability to provide the necessary social distancing in their profession for disease protection and the possibility of transmitting the SARS-CoV-2, which causes COVID-19, through aerosols and droplets formed during clinical procedures (Soysal et al., 2020). Several studies were conducted to examine its possible effects on employees and working life (in the context of remote working, motivation, digital workforce, job guarantee, and continuity). The integration of workplace and employee health rules and applications is vital in optimal management (Sezgin/Deloitte Turkey, 2020).

There are also studies suggesting that the closures applied to reduce the pandemic's spreading rate have different effects in social groups by 'gender', 'marital', and 'parental status'. As it was asserted that 'coupled women were less likely to be working than coupled men, while single women were more likely to be working than single men. Fathers of school-age children who remained employed were working reduced hours compared to men without children' (Kalenkoski & Pabilonia, 2020). A study on Turkish government employees aimed to reveal their attitudes in the context of the COVID-19 pandemic and revealed that they were adversely affected by the pandemic (Yilmaz, 2020). Studies aimed to describe the 'clinical characteristics and outcomes of older adults hospitalized with Coronavirus disease' were also conducted in Turkey (Esme et al., 2020).

Another study on the effects of the pandemic by gender regarding the fact that COVID-19 is not 'gender' oriented points to a remarkable result. The shift of resources in health services from public health to the emergency department may cause interruptions in primary health care services for women and female children, such as reproductive and sexual health services. For instance, an increase in maternal mortality due to a lack of critical resources for adolescent pregnancy was determined, and similar issues for female children not attending school were imminent. However, there may also be a vulnerability for pregnant women, whose immunity status is frequently discussed during the pandemic. Varol and Tokuc (2020) conducted a study on public health in Turkey and state that the pandemic management should be 'society-centered', based primarily on scientific knowledge without being overshadowed by economic concerns and populist policies. This study also emphasizes that the pandemic's management approach should include steps in which all humane, economic, and social, especially the social health requirements, would be satisfied within the welfare state vision. However, it should highlight that there are studies on the pandemic with a focus on immigrants. Guadagno (2020) states that with the COVID-19 pandemic's effects on access to opportunities and services, COVID-19 may trigger a migration movement from the affected areas and thereby change future pandemic patterns.

There was severe pollution, primarily through false, fake news and misleading content circulating through social networks, spreading rapidly as the WHO detailed the concept of the 'Infodemic', and drew attention to information pollution during the pandemic. The WHO, mass media cooperation, and the media's role in informing the public have also been substantial. The political dimension, which instrumentalizes the same media channels as WHO, has become another focal point in delivering accurate health-oriented information. Numerous practices under the framework of 'risk communication and community engagement' have been supportive in the WHO-led infodemic management. The pandemic clarified the need to build flexible and robust systems to deliver accurate information to communities. It has been a fundamental tendency to identify and de-mystify online-offline rumors, false information, and to take necessary measures (Golbasi & Metintas, 2020: 133). Cirrincione et al. (2020) also emphasize the problems of businesses due to a lack of knowledge regarding how and by whom information on SARS-CoV-2 is disseminated.

Specific approaches (Arslan & Karagul, 2020: 30) highlight that the concept of a 'biological weapon' is once again on the world agenda with COVID-19. Arslan and Karagul argue that humanity would experience a period in which the international community will not accept national efforts to acquire nuclear, biological, and chemical weapons after this process directly affected life on a global scale (2020: 30). The media constantly organizes actual news programs, prepares discussion programs, publishes broadcasts with expert opinions, and transfers the latest global development through international connections. Moreover, the statements made by authorized institutions were also broadcast. These are essential steps in breaking perception management. The outlets with high pollution levels by social media should not be accredited in this period, and relevant studies on this subject are a vital requirement (Kirik et al., 2020: 54).

Whitelaw (2020: 439) states that digital technologies' integration and implications are essential determinants in this era regarding pandemic policy. Countries using digital technologies 'in the race to contain the spread of a highly' contagious virus in several contexts ('planning', 'supervision', 'testing', 'contact tracing', 'quarantine', and 'clinical management') were successful in managing this burden which may lead other countries to face imminent threats.

Pandemic and Increasing Online Practices & Uses

The COVID-19 pandemic has affected the whole world considerably. Along with the pandemic, people changed their habits, and new applications began to take place at the center of human life. Personal hygiene, cyber hygiene, remote work, and online education models have been experienced as partly new situations, and people showed different adaptation levels.

In the current pandemic-driven situation, companies prefer digital technologies to obey social distance and adapt to the new normal. Despite all the new technologies, human labor has not lost its importance yet. Because, 'creativity', 'empathy', 'reasoning', 'intuition', 'interpersonal sensitivity', and 'problem solving' are *some special interpersonal human competencies that machines currently lack* (Acosta, 2020: 262).

Rules of marketing also changed with increased digitalization in the pandemic. There has been an increasing interest in e-commerce sites and banking applications. The current global crisis environment caused by the epidemic led to radical and rapid changes in consumer behavior, making consumers choose digital environments instead of physical ones. In this context, clearer understanding of the market, estimating customer expectations, and creating a social customer relationship management are consequences of digitalization in global marketing. The resulting applications inevitably attracted the attention of consumers (Marangoz & Kırlı Özen, 2021: 59). With the epidemic, the transition from physical environment to the virtual one caused consumers to acquire new online habits, and there has been a serious increase in the number of applications.

The increase in the use of social media and internet also increased the number of users of these applications. In this period, interest in Facebook, Twitter, Instagram, Youtube, TikTok applications has increased both in Turkey and all around the World, and the usage time of the applications has also extended considerably. News applications are used continuously to access daily number of COVID-19 cases, present situation, and recent developments. Individuals, spending most of their life using digital applications during the COVID-19 period, preferred video conference applications (Skype, Zoom, Meet, etc.) to attend distance education classes; and used social media and mobile applications while working from home, attending meetings, researching about pandemic, keeping up with the agenda or just to spend time, (Aldemir & Avşar, 2020: 157). This situation has inevitably caused a change in habits and preferences, and the need for technology has reached its peak.

There are extensive studies on this subject. Gemius (2020) that measured the behavior of internet users examined the use of mobile applications, which have an important place in our daily life in pandemic, in the first periods of the rise of the pandemic (April-November 2020) and in the pre-pandemic period (January 2020). According to the results, the video conferencing application Zoom was one of the most growing digital platforms of the year. The application, which served approximately 200 thousand users in January, reached over 21 million users at the end of the year. The number of Skype users increased to 12.4 million at the end of 2020, and the number of Google Hangouts users to 4.8 million. Social media application TikTok also gained momentum in this period; the number of users was 8.7 million in January 2020, while it reached 26.6 million in November. Similarly, the number of users of the music streaming platform Spotify, which was 5 million in January, increased to 17.4 million in November. The number of Netflix users, which was 4.7 million in January, reached 16.5 million in November, as one of the winners and the most talked about names of the quarantine periods.

As can be seen, COVID-19 pandemic changed digital media use habits of individuals, and new mobile applications have come into use in this period. In addition, video chat applications had a rising trend, both in business meetings and in the field of education. Thus, the transition to digital world took place very quickly. In terms of products and services, relevant data shows that the interest of users in digital applications is increasing day by day.

Distance Education and Business Models on the Basis of Sustainability

Like in many areas of life, COVID-19 had an impact in distance education and new business models. The devastating effect of COVID-19 caused many negative physical and physiological consequences, and it has been evaluated as one of the most difficult periods that humanity has experienced. Since the closures due to the pandemic have caused many economic, social, and political consequences; many fields and sectors have rapidly sought new models that aim to reconsider activism and acquisition in the pandemic process. At this point, digitality has been evaluated as an opportunity, and educational activities at many levels continued remotely and sustainably in many countries. In Turkey as well, COVID-19 epidemic affected educational institutions at all levels, and the call to 'stay at home' has made it necessary to switch to distance education applications that provide flexibility in terms of time and place. For this reason, YÖK rapidly started to labor in order not to disrupt the learning process. Comprehensive studies were carried out in five basic areas that are defined as legislation, infrastructure, human resources, content, and practice within YÖK body for distance education applications. It was determined that 123 universities have UZEM (Distance Education Center) centers and measures were taken to provide the necessary support to other universities. It was stated that within the scope of the Digital Transformation Project, a large number of students and instructors have already been trained (Genç et al., 2020: 136).

Information and communication technologies have mediated online and distance education through various software and applications so that children, young people, and adults' right to education and learning is not interrupted. Regions in which the pandemic was observed to weaken, shifted to a hybrid education model (a combiantion of face-to-face and online education systems). Distance education is the realization of learning-teaching activities, through various communication technologies, by teachers and students in different physical places, aiming to provide education services to large masses and to ensure 'equal opportunity in education'. 'With the developments in information and communication technologies, distance education model has started to have an important place in learning and teaching activities'. As face-to-face education is suspended as an emergency measure once the COVID-19 pandemic broke out, countries had to implement various distance learning applications very quickly. Among the distance education models around the world, although digital/online technology-oriented applications gained weight, there were applications where more traditional radio, television, and printed contents were used. While China uses 'Rain Classroom', an advanced and effective online platform developed in 2016, Italy makes shares through an information portal and, and in the US, the process is carried out by making 'free' and 'paid' trainings available on 'education platforms such as Edx and Coursera' (Gür Erdoğan & Ayanoğlu, 2021: 101–102).

However, many fields have experienced remote, 'working from home' model. In the course of the pandemic, employees cannot be present physically at their workplaces and started to do their jobs online with the opportunity offered by digital technologies. Although initially it was a hard period in many places, it has turned into a routine these days at which still the pandemic is unfortunately continuing. Digital environments and applications have become one of the main tools for interaction in businesses during the pandemic. In addition to change personal and social behaviors, this situation also affects the organizational structures of computer-mediated communication tools and introduces new business models. The advantages offered by digital communication networks lead to the development of new network structures that are based on communication models (Trier, 2008: 335).

Many people are worried about returning from the remote working model to the traditional one. Again, different information and data literacy levels have also affected remote working models. In terms of adaptation to this model and process, differences were also observed at the demographic level. These led to discussions about the efficiency of the remote working practice and satisfaction of the employee and the employer. In short, instead of going to the office, employees work by using information technology tools and communicate with coworkers and employers via mobile devices, phones, online chats, and videos. The impact of COVID-19 on the workforce can be seen in many sectors. This includes a change in the nature, variety, volume, pace, and value of work. Digital transformation is more than the application of a new technology. These models are also sustainable after the pandemic (Savic, 2020: 103).

Digital Security

Security, which is one of the essential needs of the individual, is a phenomenon of great importance for societies and states. Due to its importance, many security measures and precautions have been considered and used in the past and in the present day as well. Started with primitive measures in the past, the process has been technology-based today. The introduction and intense use of computers and the internet in daily life has brought a number of opportunities, while at the same time it posed elevated security risks. Many institutions, from public bodies to companies, have had to implement new-generation security measures in the digital age. Data and information confidentiality is important for various reasons. Digital security is also important at this point and includes measures against numerous cyber threats. Cybersecurity aims at best practices for protecting electronic systems against related threats. It also shows continuous improvement. Individuals, societies, and states develop various strategies and policies to protect information infrastructures as part of cybersecurity measures (Gulec & Kisman, 2021: 129–130).

With the widespread adoption of digital technologies, many aspects of society—from 'shopping and social interactions to business, industry and unfortunately crime'—have moved to the online medium. The latest reports clearly show that 'cybercrime is growing in frequency and severity'. Because, during the COVID-19 pandemic, the risks toward applications, and personal data breaches have started to increase seriously. In fact, cyber attacks have become more sophisticated and target particular victims, depending on the motivation of the attacker, such as financial gain, espionage, violence, or revenge. People had to adapt their daily routines to a new reality during the pandemic period. This situation has caused mass disruption worldwide, and the increase in working from home and social interaction, and the extensive use of digital media has brought digital security to a more important dimension (Lallie et al., 2021).

Cyber attacks are carried out for various purposes by different actors and can have devastating results. It describes an important threat in the digital world. Some advocate that cyber security, which includes 'a number of tools within itself to create and maintain the security features of the assets of institutions, organizations and users in a way that can counter security threats in the cyber environment', has become a tool of influence in international politics. Additionally, supporting data is said to exist that reveals cyber security affects the inter-state relationships. 'Cyber policies' in the context of international relations are of increasing interest (Guntay, 2018: 79–82).

A recent research examined digital data security awareness and information literacy of employees in a public institution and the effect of in-service training they receive. It was concluded that there is a moderate positive and significant relationship between the information literacy of the participants and the awareness of digital data security. However, it was determined that in-service training received by the participants did not have a significant effect on their awareness of digital data security, and level of information literacy (Avci & Arslan, 2019).

Educational, commercial, art, and even health institutions provide services to users and receive payment through online platforms. Students receive education at home, and employees of many organizations are now working remotely. Moreover, there is a sharp increase 'in demand for food deliveries and online grocery applications'. During the epidemic, almost all aspects of human life have shifted to the virtual environment, leading to an increase in 'cyber security concerns'. Cybercriminals, and others with malicious intent now see COVID-19 as an opportunity to commit cybercrime specifically for monetary gain. Many websites and applications are attacked with ransomware, and the confidentiality and integrity of data resources, for instance, patient records, face serious risk. Individuals have become victims of phishing attacks through COVID-19-related content (Gabriel et al., 2021: 275). During the pandemic period, digital security has gained greater importance.

Cyber Attacks and a Key Concept as a Set of Measures: 'Cyber Hygiene'

Rather than physical wars, the existence and effect of digital/cyber wars is a matter of concern in the digital world. These wars, which are based on fast and destructive communication technologies, cause countless negative effects and can interrupt many things instantly. For example, with a cyber attack, the websites of institutions and organizations can be crashed, social media accounts can be hacked, and access to social networks can be denied. Even, in an apocalyptic vision of the future, we can talk about such possible technological attacks as 'disrupting the energy systems of a surgery room', 'colliding trains on the subway line', and 'stopping the production lines of factories'. The concept of 'cyber hygiene' is also important at the point of these cyber attacks and is considered as a 'basic principle regarding cyber information security'. Cyber hygiene is described as 'taking simple routine measures to minimize the risks caused by cyber threats', and considered as a 'maturity model'. Cyber hygiene is also thought to be as vital as personal hygiene these days, especially during the COVID-19 pandemic at which digital/online attacks increased dramatically (Guler & Arkin, 2019: 79). We are very prone to cyber threats as we do many things online—from shopping to banking transactions, even educational and business activities. Concerns about privacy are also very important here. Increasing the level of 'information and data literacy' in order to ensure our digital security is also considered as a valid measure at this point, and this approach can increase awareness of ***digital data security*** (Yilmaz et al., 2016).

'Cyber Hygiene Campaign' conducted by the Center for Internet Security (CIS) states out some practices and issues in order to increase the basic security level of organizations. It is claimed that the practices, which can also be described as best practices in the context of cyber security, offer 'low-cost recommendations'. Therefore, they are claimed to be away from the budget and size concerns. Basic pillars of the cyber hygiene program are listed as follows (as cited in Basaran, 2015):

- *Count*
- *Configure*
- *Control*
- *Patch*
- *Repeat*

Antivirus software company ESET, on the other hand, highlighted some of the dominant cyber security skills during the European Cyber Security Month and introduced the following five basic cyber hygiene recommendations, emphasizing the importance of turning each into a cyber habit (https://www.bilgipesinde.com/internet/detay/siber-hygiene):

1. *Know the risks, make it difficult for the attackers.*
2. *Do not panic, do not rush. A few clicks can lead to far-reaching results that we do not want.*
3. *Stay updated.*
4. *Play it safe on social media.*
5. *Lock your entries.*

Organization for Security and Cooperation in Europe (OSCE) Ukraine National Project Officer Olga Voitovych stated the following about cyber hygiene (SavunmaTR, October, 15, 2021):

> Cyber hygiene is the fundamental foundation of cyber security. It is a way of thinking and a set of measures for the protection of ordinary users, public personnel or individuals in private enterprises against potential threats in cyberspace. Cyber hygiene in public administration is a necessary set of applications and measures that computer system administrators and users can install to protect and improve their online security. Similar to personal hygiene; following good cyber hygiene habits can help stay safe online. Cyber hygiene practices depend on the institutional structure of companies and organizations, so practices cover both corporate and personal behavior. The term "cyber hygiene" has many different interpretations around the world. All of the interpretations show the most important aspect of the term in relation to a particular industry...

Conclusion

In the new reality symbolized by physical distance, digital tools and processes have become more valid than ever before, and the COVID-19 pandemic has made the use of technology increasingly widespread. The pandemic has accelerated digital transformation in many areas, and companies already benefiting from the fourth industrial revolution have been able to better adapt to this process. In the current environment in which quarantine and other social distancing measures are implemented by many countries, and increasingly more people rely on the internet for information and advice, the burgeoning field for digital security has become more important. The coronavirus pandemic has also created new challenges for businesses adapting to a work model in which 'working from home' has become the 'new normal'. Companies have accelerated their digital transformation, but digital security is now a major concern.

Increase in remote working has necessitated a greater focus on cybersecurity due to heightened levels of cyber risk. Hackers use techniques to fill out credentials to gain access to confidential information of employees, and the data is then sold to other cybersecurity criminals. In fact, some of the data captured is largely due to leaks from video conferencing platforms. While the human error was already an important cause of 'cyber insecurity' before the pandemic, and employees were accessing wrong people unintentionally, the problem has grown substantially with the widespread use of technology and large-scale implementation of distance working.

Remote work and distance education have created difficulties for many small and medium-sized companies and threatened the digital security of users. The world got caught in a pandemic process that posed considerable risks to all countries. In addition, many applications have started to be used to carry out distance education and working from home activities, leading to an increase in cyber crimes. Maintaining good cyber hygiene is essential as many individuals, institutions, and businesses are now shifting to online work forms through employing a remote workforce. There are some effective measures to be taken to ensure cyber hygiene. It is extremely important to use unique and complex passwords. It is important to regularly update software installed in mobile devices and computers. It is also essential for both individuals and businesses to back up and store their data in secure environments. In addition, securing social media and e-mail accounts and using two-factor authentication are of great importance in terms of cyber hygiene.

References

Acosta, P. S. (2020). COVID-19 pandemic: Shifting digital transformation to a high-speed gear. *Information System Management, 37*(4), 260–266.

Aldemir, C., & Avşar, M. N. (2020). Pandemi Döneminde Dijital Vatandaşlık Uygulamaları. *Eurasian Journal of Social and Economic Research (EJSER), COVID-19, 2*, 148–169.

Arslan, İ, & Karagul, S. (2020). Küresel Bir Tehdit (COVID-19 Salgını) ve Değişime Yolculuk. *Üsküdar University Journal of Social Sciences, 2020*(10), 1–36. https://doi.org/10.32739/uskudarsbd.6.10.67

Avci, Ü. & Arslan, E. (2019). Dijital Veri Güvenliği Farkındalığı ve Bilgi Okuryazarlığı ile Alınan Hizmetiçi Eğitimin Etkisi. *Ankara University Journal of Faculty of Educational Sciences (JFES), 52*(3), 891–914. https://doi.org/10.30964/auebfd.493124

Basaran, A. (2015). "Siber Hijyen Kuralları", Thursday, August 13, 2015. https://www.alperbasaran.com/2015/08/siber-hijyen-kurallar.html

Bilgi Pesinde. (t.y.). "Siber Hijyen". https://www.bilgipesinde.com/internet/detay/siber-hijyen

Cirrincione, L., Plescia, F., Ledda, C., Rapisarda, V., Martorana, D., Moldovan, R. E., Theodoridou, K. & Cannizzaro, E. (2020). COVID-19 pandemic: Prevention and protection measures to be adopted at the workplace. *Sustainability, 12*, 3603. https://doi.org/10.3390/su12093603

Esme, M., Koca, M., Dikmeer, A., et al. (2020). Older adults with coronavirus disease 2019: A nationwide study in Turkey. *The Journals of Gerontology: Series A, Biological Science and Medical Science, 76*(3), e68–e75. https://doi.org/10.1093/gerona/glaa219

Gabriel, A. J., Darwsih, A., & Hassanien, A. E. (2021). Cyber security in the age of COVID-19. In *Digital transformation and emerging technologies for fighting COVID-19 pandemic: Innovative approaches* (pp. 275–295). Springer.

Gemius (2020). Mobile apps—All reports. https://www.gemius.com/all/category/mobile-apps-5738.html

Genç, S. Z., Engin, G., & Yardım, T. (2020). Pandemi (Covid-19) sürecindeki uzaktan eğitim uygulamalarına ilişkin lisansüstü öğrenci görüşleri. *Atatürk University Journal of Kazım Karabekir Education Faculty, 41*, 134–158.

Golbasi, S.D. & Metintas, S. (2020). COVID-19 Pandemisi ve İnfodemi. *ESTÜDAM Journal of Public Health*. 2020; 5 (COVID-19 Special Issue), 126–37.https://doi.org/10.35232/estudamhsd.797508

Guadagno, L. (2020). Migrants and the COVID-19 pandemic: An initial analysis. Migration Research Series N° 60. International Organization for Migration (IOM).

Gulec. Ö. & Kişman, Z. A. (2021). Uluslararası İlişkiler Açısından Siber Güvenlik ve NATO'nun Siber Güvenlik Stratejileri, *Akademik Açı, 1*(1), 127–154.

Guler, A., & Arkin, A. K. (2019). Siber Hijyenin Sağlanmasında İç Denetimin Rolü, *Denetişim*, (19), 17–40.

Guntay, V. (2018). Siber Güvenliğin Uluslararası Politikada Etki Aracına Dönüşmesi ve Uluslararası Aktörler. *Güvenlik Stratejileri, 14*(27), 79–111.

Gür Erdoğan, D., & Ayanoğlu, Ç. (2021). COVID-19 Pandemi Döneminde Eğitim Programlarının Uzaktan Eğitimde EBA Platformu Yoluyla Uygulanmasına Yönelik Öğretmen Görüşleri. *Journal of Qualitative Research in Education, 28*, 100–128. https://doi.org/10.14689/enad.28.5

Kalenkoski, C. M. & Pabilonia, S. W. (2020, July). Initial impact of the covid-19 pandemic on the employment and hours of self-employed coupled and single workers by gender and parental status. Discussion Paper Series, Iza—Institute Of Labor Economics, Iza Dp No. 13443.

Kirik, A. M., Altıntas Var, S.S., Özkocak, V. & Darıcı, S. (2020, May). Pandemi Dönemlerinde İletişim ve Algı Yönetimi: Yeni Koronavirüs Hastalığı (COVID-19) Örneği. *The Journal of Academic Social Science, 104*, 36–57. https://doi.org/10.29228/ASOS.43249

Lallie, H. S., Shepherd, L. A., Nurse, J. R., Erola, A., Epiphaniou, G., Maple, C., & Bellekens, X. (2021). Cyber security in the age of COVID-19: A timeline and analysis of cyber-crime and cyber-attacks during the pandemic. *Computers & Security, 105*, 102248.

Marangoz, M., & Kırlı Özen, E. (2021). Covid-19 Pandemi Sürecinin Farklı Alanlarda Dijitalleşmeye Etkileri: Kavramsal Bir Değerlendirme. *Hitit Journal of Economics and Politics, 1*(1), 54–68.

Savić, D. (2020). COVID-19 and work from home: Digital transformation of the workforce. *Grey Journal (TGJ), 16*(2), 101–104.

SavunmaTR. (2021, October 15). "Olga Voitovych: Siber hijyeni sağlamamak önemli zararlara sebep olabilir". https://www.savunmatr.com/ozel-haber/olga-voitovych-siber-hijyeni-saglamamak-onemli-zararlara-sebep-olabilir-h14962.html

Sezgin, C. (2020). COVID-19: Çalışanlara ve Çalışma Hayatına Olası Etkileri. March 2020, Deloitte Türkiye, Retrieved from https://www2.deloitte.com/content/dam/Deloitte/tr/Documents/human-capital/COVID-19-calisanlara-ve-calisma-hayatina-olasi-etkileri.pdf, (visited on 1 April 2020).

Soysal, F., İsler, S. Ç., Peker, İ., Akca, G., Özmeric N. & Ünsal B. (2020). The impact of COVID-19 pandemic on dentistry practices. *Klimik*, 2020; 33(1): 5–14.

Trier, M. (2008). Research note—towards dynamic visualization for understanding evolution of digital communication networks. *Information Systems Research, 19*(3), 335–350.
Van Damme, W., Dahake, R., Delamou, A., et al. (2020). The COVID-19 pandemic: Diverse contexts; different epidemics—how and why? *BMJ Global Health, 2020*(5), e003098. https://doi.org/10.1136/bmjgh-2020-003098
Varol, G. & Tokuc, B. (2020). Halk Sağlığı Boyutuyla Türkiye'de Covıd-19 Pandemisinin Değerlendirilmesi. *Namık Kemal Medical Journal*, 8(3), 579–594. https://doi.org/10.37696/nkmj.776032
Yilmaz, B. (2020). COVID-19 Pandemi Sürecinin Çalışanlar Üzerine Etkisi: Kamu Sektöründe Çalışanlar Üzerine Uygulamalı Bir Araştırma. *Third Sector Social Economic Review, 55*(3), 1724–1740. https://doi.org/10.15659/3.sektor-sosyal-ekonomi.20.08.1422
Yilmaz, E., Sahin, Y. L., & Akbulut, Y. (2016). Öğretmenlerin dijital veri güvenliği farkındalığı. *Sakarya University Journal of Education, 6*(2), 26–45.
Whitelaw, S., Mamas, M. A., Topol, E., & Van Spall, H. G. C. (2020). Applications of digital technology in COVID-19 pandemic planning and response. *Lancet Digital Health, 2020*(2), e435–e440. https://doi.org/10.1016/S2589-7500(20)30142-4

CHAPTER 6

Does E-Governance Bring New Normal Sustainable Way for Emerging Economies Post Pandemic? A Review

Eliza Sharma and John Ben Prince

Introduction

The onslaught of the COVID-19 pandemic has led to the re-evaluation of conventional systems in every sphere of society, including governance. As the world relies heavily on technology, the use of Information and Communications Technology (ICT) has become paramount in ensuring service delivery to citizens through platforms that ensure efficiency while protecting citizens from the effects of the virus. It is in this situation that e-governance has been reinvigorated to gain more traction than ever before. E-governance can be understood as the provision of government

E. Sharma (✉) · J. B. Prince
Symbiosis International (Deemed University), Symbiosis Institute of Business Management Bengaluru, Bengaluru, India
e-mail: eliza.sharma@sibm.edu.in

J. B. Prince
e-mail: john.ben@sibm.edu.in

S. R. Mondal et al. (eds.), *New Normal in Digital Enterprises*,
https://doi.org/10.1007/978-981-19-8618-5_6

services via the application of ICT tools (Tikhomirova, 2020) or the integration of several online as well as offline methods of governance (Gupta & Suri, 2017). Through e-governance, government services can be availed in an effective, transparent, and convenient fashion by the citizens (Garson, 2006). The concept first emerged during the late 1990s, with the launch of government websites and portals. According to the United Nations, the development of e-governance has continued at a great pace. The advancement of e-government is in alignment with the UN's Sustainable Development Goals for 2030. As per E-Government Survey 2020, the global average EGDI (E-Government Development Index) value has increased to 0.60 in 2020, from 0.55 two years ago. The numbers are more interesting in the case of developing countries, where countries with low-to-middle levels of income have seen a 57% rise in digital governance (UN E-Governance Report, 2020).

There are four pillars to e-governance, namely; connectivity, knowledge, data, and capital (Yadav & Singh, 2013). Harmony between these four factors is necessary to achieve efficiency in e-governance. The beneficiaries of e-governance can be other governments, public employees, businesses, or the citizenry as a whole. Public officials use e-governance tools for maintaining workplace records such as contact details of employees or to file complaints against system crashes and faults (Bala & Verma, 2018; Baroi & Alam, 2021). Government agencies can use technology for database sharing or e-accounting to rapidly transfer government documents or other information between various departments or states (Manoharan et al., 2021). Similarly, businesses can also avail services such as information regarding updated industrial guidelines and tax collection processes. The digital process allows the businesses to move past cumbersome paperwork and red-tapism, thus streamlining the bureaucratic process and dodging corruption (Ray & Mukherjee, 2007). Lastly, the citizenry as a whole can use government websites and online portals to take advantage of services such as administration certificates, information-availability, and online bill payments. The applications of e-governance can be seen across multiple sectors-health, education, banking, or administration (Bhanti et al., 2011; Ray & Mukherjee, 2007; Yadav & Singh, 2013).

The primary goal of e-governance across all sectors of application is to reach the citizens or beneficiaries and to ensure that their service requirements are well- met. Indeed, the concept of e-governance is essentially a step towards SMART governance, which implies, Simplicity, Morality,

Accountability, Responsiveness, and Transparency (Lee-Geiller & Lee, 2019). The advantages of e-governance thus lie in the improvement of government interactions with the public and industry. It enhances the efficiency of services, leverages technology as an enabler (Mohammad Ebrahimzadeh Sepasgozar et al., 2020), and leads to cost reductions and lesser corruption in the bureaucracy. Due to the flattening of the conventional hierarchical organizational structure, it also lends more legitimacy and transparency to the administration, thus improving the relationship between the public and the public authorities (Batta et al., 2012). Amidst the ongoing pandemic, these advantages of e-governance become more important and impactful to ensure citizen safety and logistical efficiency in governing during these turbulent times.

E-Governance and COVID-19

Despite the great progress in ICT and the private sector, the public sector has lagged in digitizing its operations around the globe, barring a few exceptions. E-governance also forms part of the Sustainable Development Goals 2030 decided by the UN. According to the UN, e-governance report UN E-Governance Report (2020), the digital shift proves the efforts of countries towards providing effective, sustainable, and accountable services to their citizenry and bridge a digital divide between the people and the government. However, with long queues, mountains of paperwork, and close-contact appointments, the age-old setup of governance and bureaucracy functions was essential to be transformed in the wake of the COVID-19 pandemic. The urgent necessity to limit all possible avenues of contact and exposure to the virus had pushed governments sharply towards the wide adoption of e-governance (Bajpai & Biberman, 2020). Great preference was given to remote-working or telecommunications and shifting basic administration services to online portals. Quarantine and lockdown measures around the world, along with social distancing policies, have raised questions about how digital governance solutions may be used to ensure the consistent fulfilment of civil services (Asongu et al., 2021) while ensuring security as well as accessibility. The pandemic has also led to a fresh wave of fake, unreliable, and even dangerous, misinformation on the internet, which needs to be countered with reliable and authenticated information about the virus, to avoid panic responses among people (Gupta et al., 2020; Shahi et al., 2021). There have been several new ways in which governments have adopted e-governance, namely; increasing digital capacities, providing regular and updated data, coordinating aid efforts, and creating online portals as a way to engage and interact with the citizenry rather than a singular bot (auto-robot) or video (Batta et al., 2012; Mittal & Kaur, 2013; Ullah et al., 2021). Several governments have also decided to take the hybrid approach, which involves the integration of online and offline multichannel delivery of essential goods and services, and the expansion of e-participation by citizens for a fraction of the process. Developing countries additionally benefit from the applications of e-governance as it provides equal platform to all the citizens (Bala & Verma, 2018) while creating greater accountability of government services (Banerjee et al., 2020). Citizens living in remote areas or belonging to underprivileged communities can gain easy access to services that are otherwise time and money consuming. In the absence of face-to-face communications, information delivery is important; countries have used versatile ways to provide clear, updated information to the citizens (Addo & Senyo, 2021). It is hence crucial to see how various countries, whether developed or developing, have utilized the advantages offered by e-governance during COVID-19.

Research Design

The research is structured thus: Firstly, an overview of the policies across different countries are listed with brief explanations. This reveals key information such as communication modes, the type of information, and how electronic means were used to share vital data with citizens and enable easy transactions—both monetary and non-monetary. Following this, we study the impact of ground-level initiatives or implementation and its influence on COVID-19 in select countries. The practice perspective helps understand specific actions and how they helped reduce the spread of the pandemic. The insights and facts are represented in a tabular form with due references. Subsequently, details about how different stakeholders such as community, general public, and healthcare workers were affected in varied ways are explained. We note down efforts taken through e-governance mechanisms and adopted by different countries to tackle various issues. Following through, the next section deals with recommended policy and action measures in the form of a framework specific to

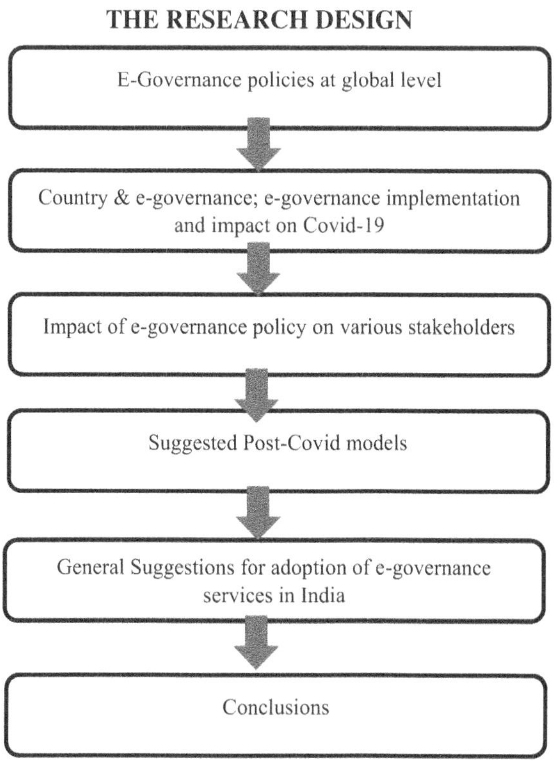

Fig. 6.1 Research design

e-governance dealing with pandemics. Lastly, we lay down policy implications for practitioners, institutional functionaries, and governments. The research design is shown in Fig. 6.1.

E-Governance Policies at the Global Level During COVID-19

The popularity of e-governance stretches across the globe. Many countries have woken up to the advantages of e-governance as it shows a more efficient, responsive, and transparent approach for the delivery of citizen-centric services in their respective countries. As countries strive to control the spread of COVID-19 while facing the challenge of running high-stakes services to citizens, technology has lent a helping hand (Elavarasan & Pugazhendhi, 2020; Melitski, 2019). The following table presents examples of e-governance practices adopted by developed and developing countries around the world to tackle the challenges posed by COVID-19. Table 6.1 gives an overview of practices in e-governance across various countries. Table 6.2 gives an overview of ground-level implementation in different countries and its impact on COVID-19.

Impact of E-Governance Policies on Various Stakeholders

1. **Society or Community**

The governments around the world have launched several initiatives to ensure that the impact of COVID-19 does not deter good governance from reaching the community. In Pakistan and Sri Lanka, where the bulk of the daily economy runs in the unorganized or informal sector, despite stringently imposed lockdown, the governments have issued e-Passes for vendors, hawkers, supply chain workers, essential workers such as sanitary and hygiene workers for ease of movement. In India, the government launched a new Twitter handle, named *@CovidIndiaSeva*, to provide transparent, real-time delivery of services and swiftly answer citizen's questions. The Twitter handle creates resolvable tickets from citizen queries on the platform, and then forwards the same to the concerned authorities for resolution, often conveying subtle sentiments along with concrete facts (Arora et al., 2021). The Ministry of Technology also issued a virtual diagnostic app called Arogya Setu, which could help to calculate the potential risk of getting infected. The app is available in 11 languages; it was downloaded more than 150 million times within two weeks of its launch. In South Korea and Indonesia, the early release of technical guidelines on online platforms regarding the prevention, precaution, control, isolation, and hospitalization aspect of the COVID situation helped to manage the situation and avoid panic and chaos in the public. Similarly, in Canada and New Zealand, Prime ministers and key government functionaries regularly address the public (New Zealand Herald, 2022) during the announcement of strict lockdown measures. This helped spread the right information and awareness about the pandemic.

Table 6.1 Country & e-governance practice

Countries	*E-governance practices*
Argentina	1. @TeDis—Assists persons with disabilities. It's an e-service that aims to make E-governance services more accessible to everyone 2. Educ.ar is Argentina's education portal. It functions as a virtual platform for connection and contact with the country's educational department (Education Argentina Portal, 2022) 3. 'Zero Balance' is a regional education initiative aimed at lowering CO_2 emissions and raising environmental awareness among young children
Austria	1. A national eID solution—Allows citizens to identify themselves to e-governance services in a unique way (Austrian Government Portal, 2022) 2. Citizens have an option of generating electronic signatures for e-governance procedures 3. Allows citizens to address tax-related matters online 4. Assists residents in filing claims for social security benefits such as unemployment benefit
Chile	1. Conversion to digital Municipal is a part of municipal digital development and integration. It aims at making municipal governments accessible to inhabitants without requiring them to relocate from rural to urban areas 2. The Electronic Medical Leave Certificate (EMLC)—For employees to file medical leave forms electronically rather than on paper, thanks to the solution 3. Memoria Chilena—A gateway that provides access to National Library of Chile's resources. It features books, newspaper articles, images, video assets, and documents that shape Chile's national identity, promoting further research (Chilena, 2013)

(continued)

Table 6.1 (continued)

Countries	*E-governance practices*
Columbia	1. Agronet—A platform that provides strategic information to agricultural transaction agents, such as policymakers and small farmers. It maps and analyzes land use across municipalities using Geographic Information Systems (Gogo et al., 2014) 2. Colombia's Plan Vive Digital is a road map for government technology intervention. It aims at enabling Internet access to every resident, as well as the establishment of a national digital ecosystem
Germany	1. A national eID solution—Allows citizens to identify themselves to e-governance services in a unique way (Noack & Kubicek, 2010) 2. Allows citizens to complete tax-related tasks online 3. Citizens can generate electronic signatures during e-governance procedures 4. Assists residents in filing claims for social security benefits such as unemployment benefit
Kenya	1. The system delivers information on different technologies of farming, agricultural markets, types of business models, and advice to its users 2. Free registration is available for farmers and agricultural product dealers in the distribution system. By providing your name and phone number, you may quickly register on the website or via SMS. They will get access to all farming and product information once they have completed the registration process. Farmers can also use the system to submit comments on agricultural products and locate the nearest retailer location 3. The Budget Tracking Tool—For communities to actively engage in public resource management and to enable citizens to monitor and track the movement of development funds 4. Bunge SMS—It aims to empower the citizens to influence local governance in their constituencies. People can report any sort of violence, corruption, environmental degradation and also monitor development activities (Baguma & Eilu, 2015)
Mexico	1. @Campus Mexico—To provide learning opportunities and training to assist government servants 2. Seismic Early Warning System (SAS)—Along with the system, there is an application that will send out an alarm roughly one minute before the quake occurs. The Centro de Instrumentación y Registro Ssmico (CIRES), a civic association that coordinates Mexico's Seismic Early Warning System, is directly linked to the Blackberry application (Suárez et al., 2018)

Countries	*E-governance practices*
Myanmar	The Asian Development Bank (ADB)—To find strategies to develop government applications and online services, and to equip them with human and institutional capital (Simpson & Park, 2013)
Netherlands	1. A national eID solution—Allows citizens to identify themselves to e-governance services in a unique way (Cuijpers & Schroers, 2014) 2. Allows citizens to complete tax-related tasks online
Oman	1. The AL-SHIFA programme—For medical practitioners to maintain record of a patient's medical history, provide proper treatment and followup and communicate information with other experts as required 2. The Land Applications Service—For residents and enterprises. Citizens and businesses can apply for land if they meet certain criteria 3. The e-Police programme—This is Oman's first electronic service with e-payment capacity. The multilingual Portal provides various benefits to all the people residing in Oman, including access to services such as e-payment of traffic tickets, online visa application, visa status inquiry, etc. (Royal Oman Police, 2022)
Pakistan	1. Pakistan Digital—For the country's e-commerce and electronic government services 2. Common government services available online include downloading forms, obtaining information, using job services, searching for library books, submitting forms, making payments to various government entities, interacting with the tax office, requesting documents, completing change of address forms, police statements, and car registration (Malik et al., 2016) 3. Citizens can also use the website to pay parking tickets, request for passports, driver licences, birth certificates, and other replacement documents
Saudi Arabia	1. The SaudiEDI programme—For managers to follow and monitor import and export procedures in order to improve transparency and public service 2. Government Service Bus (GSB), the Government Secure Network (GSN), and the Saudi Government Portal—For everyone in the kingdom to have access to governance services from anywhere and at any time (Alannsary & Hausawi, 2019)
Spain	1. A national eID solution—Allows citizens to identify themselves to e-governance services in a unique way 2. Citizens have an option of generating electronic signatures for e-governance procedures (Arteaga & Criado, 2012) 3. Allows citizens to address tax-related matters online 4. Assists residents in filing claims for social security benefits such as unemployment benefit

(continued)

Table 6.1 (continued)

Countries	*E-governance practices*
Thailand	1. Created a unified digital infrastructure, including the Government Information Network, in order to improve communication and sharing information with intra- and inter-agency 2. The ministry is in charge of digital legislation and cyber security, as well as initiatives for using technology to engage citizens in policy-making processes 3. The Electronic Government Agency is making developments in data technology solutions—To increase the efficiency of access to data across public agencies for improvements. This involves data integration and applications to help with big concerns like human trafficking, illicit fishing, access to health care, and education 4. Government Smart Kiosk—Online platform for doing business, GovChannel and GNews—To allow simple access to essential information for those dealing with public service transactions (Kuzma, 2010)
Turkey	1. A national eID solution—Allows citizens to identify themselves to e-governance services in a unique way (Mutlugün & Adalier, 2009) 2. Citizens have an option of generating electronic signatures for e-governance procedures 3. Allows citizens to address tax-related matters over the internet 4. Assists residents in filing claims for social security benefits such as unemployment benefits
United Arab Emirates	1. Department of Smart Government (DSG) of Dubai—Aims at providing Electronic Shared Services (ESS) to residents of UAE especially the professionals. People's ability to offer comments and contributions to government policy has tremendously aided their creative and imaginative thinking (Alketbi, 2018) 2. The UAE government has established the UAE ID programme, which will give people cutting-edge multipurpose smart ID cards
United Kingdom	1. A national eID solution—Allows citizens to identify themselves to e-governance services in a unique way (Bruegger et al., 2007) 2. Allows citizens to complete tax-related tasks online 3. Assists residents in filing claims for social security benefits such as unemployment benefits 4. Allows citizens to get photocopies of registry office certificates such as birth, marriage, death certificates

Table 6.2 E-governance implementation and its influence on COVID-19

S No	*Country*	*E-governance implementation*	*Impact on COVID-19*
1	Greece[1]	• Lockdown-passes issued through special SMS facility • New e-platform set up to give virtual access to public services	Nearly 110 million SMS sent free of cost during March–May 2020 The e-platform allowed thousands of high-risk individuals to access important documents
2	Estonia[2]	• Nearly 99% of the government services provided via the e-Estonia project • Introduction of a chat-bot named Suve, to make information-sharing interactive • Workforce-collaboration platforms	Transparent and accessible information to the citizenry leads to a better understanding of safety rules Collaborative platforms and efforts ensured the availability of work and ease-of-information-access
3	Britain[3]	• British Parliament proceedings held via video-conferencing • Cameras and AI algorithms used for traffic control deployed for COVID- policies	Allowed governance and legislation to continue uninterrupted during the pandemic The cameras and sensors are used to judge the distance between two people to warn about social distancing
4	Denmark[4]	• National contact tracing app (COVIDmeter) • Mobile Proximity App featuring self-assessment questionnaire and self-reporting systems • Digital post services	To detect sustained contact within 1–2 m with potential virus carriers; Health authorities will only be able to use data from the app to follow up on the spread of the infection

(continued)

[1] Government Website, www.gov.gr.

[2] Estonia Government website, https://e-estonia.com/.

[3] United Kingdom Government website, https://www.notifications.service.gov.uk/.

[4] Denmark, https://lifeindenmark.borger.dk/.

Table 6.2 (continued)

S No	*Country*	*E-governance implementation*	*Impact on COVID-19*
5	Republic of Korea[5]	• Use of AI to develop a quick COVID testing kit, and use of digital bureaucracy to give it quick approval from authorities • Smart quarantine information systems, Contact tracing systems • AI-enabled diagnosis systems	The AI-enabled approach led to the rapid development of COVID testing kits and similar kits then developed around the country; comprehensive and efficient info systems to help frontline health workers about suspected COVID cases; Use of AI for better diagnosis and patient classification
6	Finland[6]	• Authorities enlisted social media influencers and celebrities to spread COVID information via MediaPool • Daily streaming of information sessions by Mayor of Helsinki • Digital cultural services for stimulating normalcy in times of lockdown • Personalized, doorstop services to elderly and at-risk individuals	Enables govt authorities to inform people outside traditional media network; ensures rapid deliverability of reliable information in a crisis; Consistent updates on government response and policies; Mitigating the psychological impact of lockdowns, especially upon the aged populace
7	Argentina[7]	• ANSES, Argentina's social protection agency, gave an immediate emergency funding to informal sector workers • 2000 digital governance services provided online • Remote digital ID processing	More than 8 million informal sector workers impacted directly through digital money transfers by the government 90% of the population had internet access, which made govt services nationwide
8	Canada[8]	• COVID alert app • Web-based self-assessment tool • Mental health portal • Email Updates regarding COVID-19 news and information	Notification and updates about positive COVID patients in the immediate area; mental health and substance abuse portals for people at risk; subsidy calculators for financial aid to citizens

[5] South Korea, http://www.korea.net/.

[6] Finland website, https://valtioneuvosto.fi/en/frontpage.

[7] Argentina Government website, https://www.argentina.gob.ar/.

[8] Canada Government website, https://www.canada.ca/en/government/system/digital-government.html.

S No	*Country*	*E-governance implementation*	*Impact on COVID-19*
		• Wage Subsidy and other financial aid calculators	
9	Italy[9]	• Self-isolation warnings and reminders sent to suspected cases via SMS • Digital solidarity site that offered free access to newspapers and magazines, e-learning tools, cloud computing	The SMS facility has led to early isolation practised by potential COVID cases, or availing medical aid in a timely way; helped citizens to deal with the psychological and economic impact of pandemic and lockdown
10	USA[10]	• Made Unemployment Insurance System be more robust • Volunteer-matching systems for medical aid, food distribution, technical experts • COVID information portal in multiple languages	The benefits application process ran smoothly, allowing people to avail insurance money when out of work More than 1000 experts from different fields matched with their required vacancies across 29 states
11	India[11]	• Multilingual and state-specific apps like Arogya Setu and COVA launched to give essential services and information • One-step platforms like CovidTracker and E-Sevai for reliable news dissemination and relevant information • Online issuance of Citizen/service provider passes during the lockdown • Virtual education through public-owned channels like Doordarshan	Applications allowed self-assessment self-reporting, services like travel Pass and groceries to millions of people; allowing citizens to mark themselves as 'quarantine' and inform others; regular govt updates and push notifications to keep people well-informed. Up to 150 government services are offered online. Also Ensured the continuity of education to an estimated 75 lakh students despite lockdown and closing of schools

[9] Italy Government website, https://solidarietadigitale.agid.gov.it/#/.

[10] United States of America website, https://www.usdigitalresponse.org/.

[11] Indian Government websites https://mygov.in/; https://mygov.in/aarogya-setu-app/.

2. **Migrant Students and Workers**

The suddenly-imposed lockdowns during the COVID pandemic saw a large-scale exodus of migrant workers across the world. In India, the situation was similar with the vast majority of migrants moving from metropolitan cities to their hometowns, disrupting the revival of business activity post-lockdown (Khanna, 2020). The governments grappled with the logistics of this unprecedented situation by creating online databases, which in India was named *National Migrant Information Survey*—for the benefit of the returning or travelling migrant workers and students in the country. These databases were aimed at helping states as well as migrants themselves to monitor their movement and also to find help in travelling to their homeland. The lists of special trains allocated for migrant journeys- Shramik Trains- were also updated in the database to provide clear information about the changes in schedule or any other important information. Per official estimates by the Central Information Commission (The Hindu, 2020) these measures would benefit at least 26 lakh migrants currently stranded across the country in halfway houses. In the United States, updated travel and trade advisories have been regularly issued on the website of Customs and Border Patrol for the benefit of the travellers. In Japan, where many workers and students temporarily relocate to Tokyo for work and education, the mapping of migrant routes and current locations on the country map also sought to link their bank accounts, to provide government subsidies and emergency aid to them via direct transfers. In countries with federal governance structures like Canada and India, cooperation between different states, headed by the central government, ensured that there was better coordination, movement monitoring as well as contact tracing in this case.

3. **International Travellers**

For the sake of international travellers, the governments have provided facilities that catered to their travel and visa needs, lodging and accommodation issues, and information availability regarding their application status, among other things. Several countries under the European Union, such as France, Germany, and the Netherlands, adopted new air-travel restrictions and preventive measures. These governments also provided a way to provide for faceless and contactless transactions such as e-ticketing and web check-in for these travellers via an online portal, which carried self-assessment forms and self-declaration documents. The objective of e-registration was to avoid hurdles and inconveniences to foreign travellers due to overcrowding and queues at airports and border checkpoints. Travellers were given the option to self-register on web portals from right their homes, as per all guidelines and checklists issued by the government, and could also ensure hassle-free travel. In the United States, regularly updated traveller guidelines were uploaded by the Department of State and U.S. Citizenship and Immigration Service (USCIS) on the official websites (Chua et al., 2020) as more information about COVID prevention and precautions came through. Travel advisories were not just uploaded on the websites, but also sent via emails and SMS to travellers, featuring detailed instructions about using masks, protective equipment, and necessary etiquette during their trip. Government announcements and press releases also pointed out the need for a self-isolation period of 14 days during COVID-19 after a trip from abroad.

4. **Domestic Travellers**

For domestic travellers, the governments in many countries with federal structures announced mandatory re-registration before every inter-state movement and intercity movement in designated hotspot areas. In Canada, the process of e-registration was made smoother and quicker. In Italy, the traveller had the option of printing the acknowledgement of the pass and later, sticking it to his car window for a paperless check of documents. Further, officers were also taught about basic medical screening, and about asymptomatic carriers of the virus. In Italy, Japan, and Korea, the monitoring of these travellers was done by tactics such as CCTV traffic cams, self-imposed quarantine for at least 14 days (Korea Times, 2021) after their return. In India, travellers have also witnessed the benefits of m-governance, which is governance through mobile facilities, including SMS alerts regarding health and safety guidelines. These messages also helped the travellers through the wide circulation of emergency contact numbers for medical aid, information, law, and order, or quarantine and isolation details. By the use of travel details through mobile phone usage, the governments could pinpoint the citizens who had travelled during the lockdown period and issue special notices to them in addition to mass alerts. Measures like contact tracing through technology enabled the identification of potentially at-risk travellers by the concerned authorities in India, which aided them to alert the individuals and advise them for self-isolation.

5. **Businessmen/Vendors**

The government had to ensure that the industrial downturn does not sustain for a long per cent for businesses and vendors. To ensure this, apps were launched at hyperlocal levels by state governments in the USA, Bangladesh, and Canada. These kinds of apps had the following features; (a) receipt of requests from households for essential items in containment zones, it was then proceeded by real-time delivery tracking; (b) management and other logistics of compiling a set of vendors for door-step deliveries—including essential items like groceries and fruits to ensure there was limited footfall and adequate social distancing at the local markets; (c) allocation of mandatory and stringent time slots to citizens for purchasing needs. Additionally, the local governments in India have also been quite forward in innovation and leveraged these solutions to provide livelihoods in their locations. A system was devised across the country where hand cart hawkers, street vendors, or rickshaw pullers were paid for groceries on a daily/weekly basis to make door-step delivery to customers in a particular locality. Moreover, the customers also directly paid through online channels like Paytm to a local shop without needing to physically visit the shops, thereby maintaining contactless purchases.

6. **COVID Patients**

The prevailing COVID-19 situation meant that every suspected, as well as confirmed case of coronavirus, had to be detected and isolated to fully strengthen any containment measures. This was done through contact tracing by dedicated IT teams drafted by state governments for this purpose. In India, the use of self-assessment forms on web portals like Aarogya Setu which integrated with mobile applications (Goyal et al., 2021) also helped the government's efforts by involving the honest participation of the citizenry. The awareness for similar portals was also spread via the use of mass media such as radio, television, SMS alerts, and social media accounts of the state and central government in tech-friendly countries like South Korea, Japan, New Zealand, and Finland.

7. **Social Workers**

Social workers such as sanitation workers were as much at the forefront of the pandemic as health professionals. To ensure their safety and ensure their spirit of service, several measures were taken under digital governance. These included timely updates regarding required safety precautions and guidelines about protective clothing and masks. The government of Singapore and Denmark also used special AI-enabled online portals to set and monitor specific routes taken by sanitary workers, and carefully arrange them according to hotspot areas. In India, social workers were routinely offered free testing using the gold standard for testing—the RT-PCR test. Frequent updates were provided to them using Aarogya Setu, the government's e-portal and digital app. Moreover, they were given top priority when the first round of vaccinations commenced in early 2021. The measures were undertaken to not only take care of infections but also provide them with a sense of assurance from the government.

8. **Doctors and Hospital Staff**

Certain common features of healthcare and medical management-related apps and portals introduced all around the world to combat the COVID-19 rush that is being used include (a) tracking the geographical spread of the pandemic on a real-time basis using GIS-linked tools; (b) features for the real-time patient and related data upload by healthcare professionals, municipal officials, COVID-19 call centre operators into a single central database; (c) monitoring of availability and utilization levels of hospital facilities and beds; (d) managing the demand, availability and utilization levels of equipment like ventilators as well as essential medical items, including N95 masks and PPEs. In India, the mass awareness media channels clarified that it was mandatory for all hospitals (Government and Private), Medical officers in Government health institutions, and registered practitioners to also get the self-declaration forms, who were having a travel history of COVID-19 affected countries or treating COVID-19 patients as their duties. Steps like e-token and telemedicine introduced by various governments have also reduced the burden upon doctors and reduced the chances of secondary infections from hospitals and clinics. Further, the governments in all countries have invested in tech that shields doctors, such as Aerosol boxes and mobile testing booths, that keep doctors and medical professionals secure from being infected themselves by the virus as they battle on the frontlines. In Australia and the USA, telehealth technologies were put to full-scale use through robots, drones, and technologies that can perform functions such as disinfection or remote patient monitoring. In China, robots were deployed for monitoring the health status of the patients, delivering meals, and disinfecting the surfaces.

Suggested Post-COVID Models for Development Using E-Governance

In this scenario, it is essential to look at the possible models and suggestions that may be adopted for the successful development of digital governance policies. Few relevant models for e-governance have been discussed below, which can be considered as a base model for effective and efficient planning, implementation, and execution of e-governance policies by the governments in affected countries.

The Layered Framework for Implementation of E-governance services has been adapted from Das et al. (2014) represented as Fig. 6.2. to combine the functionalities and operations for the effective implementation of e-governance services. This model has been studied in order to understand the various tiers that make up the framework of any e-governance portal, in order to provide an effective e-governance delivery to the citizens. The model presents layers of e-governance implementation, prioritized as per its importance. The client layer represents the delivery channels of e-governance. The second layer is the presentation layer, which represents the interface of interaction between the client and the government. The third layer, similarly, is the application layer, which focuses on the core of the strategy and lodges the web services of the administration. The services layer back-end operations that perform authentication and forward operations. The fifth layer includes databases for storage and persistence of data. The given model is cost-effective and strategic for e-governance implementation and can be suitable for constructing a framework of digital adoption in Indian governance.

The E-Governance Cloud Computing Model has been adapted, taking cues from Mukherjee and Sahoo (2010) to implement cloud computing technology with e-governance. The representation is outlined in Fig. 6.3. Cloud computing is a business model of delivering IT resources and

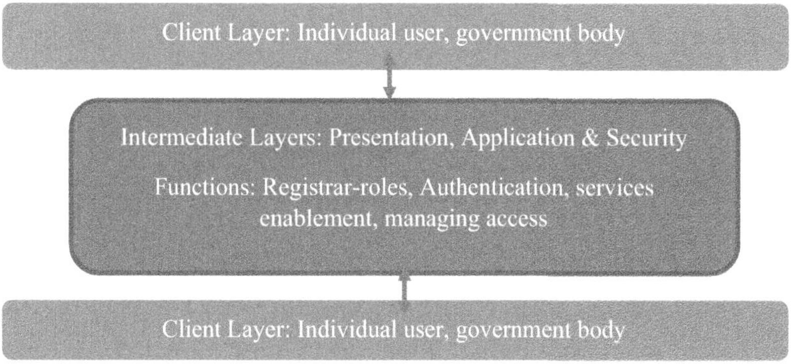

Fig. 6.2 Layered framework for implementation of e-governance services

applications as services accessible remotely over the Internet rather than locally. This model has been studied to understand the extent of new technology implementation into e-governance systems, rather than simply relying upon hybrid systems such as e-Pass or delivery apps (Giri et al., 2018). Per this model, cloud computing can provide a cost-efficient and data-efficient framework for e-governance, which includes Infrastructure as a service (IaaS), along with Platform as a Service (PaaS), and Software as a Service (SaaS). Shukla and Mathur (2020) emphasize the role of leveraging data stored on the cloud and also recommended analytics to sharpen performance of e-governance. Moreover, the benefit of using cloud computing models is that it can handle large volume of data and transactions, which are required in a country with large population such as India and Russia. It thus promotes efficient infrastructure, platform, and service compatibility along with excellent data integrity and data recovery options, which are needed in situations where digital literacy is low and the chances of mistakes are rampant.

The discussion of the above models provides insight into suggested structures for adopting and developing e-governance facilities in a worldwide context. At the same time, however, some general suggestions may also be noted in order to ensure ease of implementation and efficient results during the adoption of these models or others based on their insights. These general suggestions will be beneficial to policymakers, researchers, corporate, and administrators in developing e-governance solutions that work according to the challenges present in countries

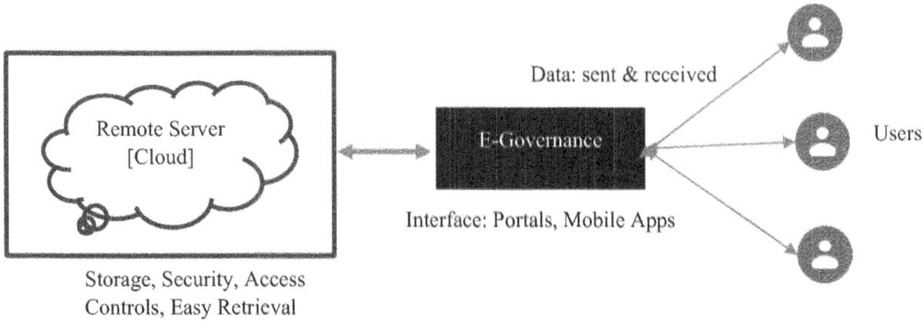

Fig. 6.3 E-governance cloud computing model

that have yet to embrace e-governance policies in full measure. These suggestions have been presented below.

General Suggestions for the Successful Adoption of E-Governance Services in Indian Society

1. ***Overcome Language Barriers***
   Despite the immense influence of modernization and globalization, multilingual identities thrive in several countries around the globe. There can be a rich cultural heritage of different languages and regional accents, which can nevertheless present a challenge when it comes to creating governance platforms (Karmakar, 2020) that are common to all citizens of a country. The Indian context presents unique challenges in the form of tribals who live outside routine habitats such as cities, towns, and villages, thereby causing considerable hurdles to the effective implementation of e-governance. This is due to a lack of access and inability to communicate in their local dialects (Chaudhary, 2020). In this case, the diversity of languages must be kept in mind while designing efficient government websites and systems, with options to access local language translated versions of these citizen portals (Pawar et al., 2016). Software vendors should be encouraged to instal local language support, including in user tutorials and online guides to the services.
2. ***Citizen Awareness Programmes***
   Along with low literacy, several developing and underdeveloped countries in the world also suffer from IT or digital literacy. To counter this issue, governments must ensure that their e-governance

measures are simplistic and easy-to-learn for even the most digitally-backward communities in their countries (Buchholz et al., 2020). Marketing and publicity efforts should be made for creating awareness regarding the availability of government services through digital platforms. IT literacy programs should be offered early in schools for the benefit of the next generation (Mittal & Kaur, 2013). Further, there must be online as well as offline well-trained citizen-service centres (such as Suvidha Kendras in India) to address the issues encountered by people with little to no digital education.

3. ***Enhanced Connectivity***
   A major challenge can manifest in the countries where large portions of the population live in rural or semi-urban areas, where network connectivity is low. This lack of infrastructure must be tackled by providing easy internet access outside cities and concentrated urban areas. The governments may provide an incentive to Internet service providers, and allot special budgets for rural internet infrastructure (Pawar et al., 2016; Rahman, 2010). The use of mobile internet vans, or common wi-fi hubs, and shared IT labs can be set up for encouraging the people to employ technology in their daily lives.
4. ***Research and Implement Best Practices***
   Countries that are seeking to implement e-governance policies must invest in building research bases where the best e-governance practices followed around the globe can be studied and then developed according to the specific context in that country. This reduces the struggle of trial-and-error, and the practices of developed countries can inspire their adoption in the developing countries. In this chapter, several e-governance practices during COVID-19 by developed countries like Estonia, USA, and the UK have been highlighted, which could be implemented by these governments as till today the cases of COVID-19 cases are increasing at a global scale and the graph has not flattened.
5. ***Usability Surveys and Feedback Loops***
   Governments must institute easy and incentivized feedback systems in their e-governance projects to gather useful data regarding how to improve the effectiveness of these systems in their countries' unique scenarios especially in times of natural disasters and pandemics (Janssen & Van Der Voort, 2020). The governments must not be complacent and establish easy systems without necessary follow-ups to ensure their effectiveness. Strong, critical feedback is necessary

to build a lasting rapport with the citizenry and overcome the massive divergence in levels of implementation of e-governance among the different states. E-readiness surveys and quality checks must be carried out across all such initiatives and their lessons and insights are emulated across the board (Bhanti et al., 2011; Yadav & Singh, 2013). Feedback may also be collected through observations, comments, official notes that may provide information regarding e-governance application, and citizen satisfaction.

Conclusion

E-governance projects can play an essential role in improving public service and value. The objective of e-governance projects must be to improve the service delivery of citizen-centric projects. Many countries around the globe still grapple with challenges such as low digital literacy, poverty, and lack of awareness; it becomes difficult for e-governance to achieve a breakthrough in its complete installation and implementation. However, over the past few years, there has been major digital outreach on the part of various governments, especially in developing countries, as they adapt to serve a more digitalized citizenry. In the wake of the COVID-19 pandemic, the scope of digital services has been realized furthermore, with the governments heavily-reliant on using technology and remote access to databases for planning, implementation, and coordination of government services and policies. This chapter has focused on providing examples of COVID-19-specific e-governance policies adopted in developed as well as developing countries across the world. There is a huge scope in the advancement of e-governance services, especially in developing and underdeveloped countries, post the COVID-19 pandemic. Going forward, governments need to develop e-governance as a necessary core competence that facilitates implementation at the ground level (Hooda & Singla, 2020). The massive and favourable response to e-governance measures like dedicated mobile applications, e-Pass or e-permission facilities, aggrieved citizens' databases et cetera has shown that the citizens of these countries today are eager to develop e-readiness and willingness to shift to e-governance models for quick and effective services. Indeed, a transparent practice of e-governance can have a direct influence on the information level of citizens and empower them. Additionally, the efficiency of e-governance over traditional methods of administration leads to increased productivity, which boosts the country's financial and industrial health as well. Thus, e-governance leads to a stronger society as well as a stronger economy (Rahman, 2010). In the era of COVID-19, we have found that the services provided by successfully adopted e-governance can even end up saving lives. Certainly then, these areas need to be focused upon immediately and urgently. Hence, e-governance is an interesting field of research for further study and holds immense potential for advancement. At last, general suggestions for the effective adoption of e-governance in less-equipped countries have been mentioned; based on the study of barriers which has hampered the maximum utilization of e-governance policies of these governments.

References

Addo, A., & Senyo, P. K. (2021). Advancing e-governance for development: Digital identification and its link to socioeconomic inclusion. *Government Information Quarterly, 38*(2), 101568.

Alannsary, M. O., & Hausawi, Y. M. (2019, March). Adopting and implementing a government cloud in Saudi Arabia, an integral part of vision 2030. In *Proceedings of 34th International Conference on Computers and their Applications*, *58*, 387–396.

Alketbi, H. (2018). *An evaluation of e-government effectiveness in Dubai smart government departments* (Doctoral dissertation, Southampton Solent University).

Arora, A., Chakraborty, P., Bhatia, M. P. S., & Mittal, P. (2021). Role of emotion in excessive use of Twitter during COVID-19 imposed lockdown in India. *Journal of Technology in Behavioral Science, 6*(2), 370–377.

Arteaga, S., & Criado, I. (2012, June). From technology diffusion to social use: The case of the eID card in Spain. In *Proceedings of the 12th European Conference on eGovernment* (p. 66).

Asongu, S. A., Diop, S., & Nnanna, J. (2021). The geography of the effectiveness and consequences of COVID-19 measures: Global evidence. *Journal of Public Affairs, 21*(4), e2483.

Austrian Government Portal. (2022). *Electronic identity (eID) of other EU member states*. https://www.oesterreich.gv.at/en/themen/dokumente_und_recht/elektronische-identit%C3%A4t-(eiD)-anderer-eu-mitgliedstaaten-(SDG).html#:~:text=The%20central%20eIDAS%20node%20of,taking%20place%20in%20the%20EU%20

Baguma, R., & Eilu, E. (2015). Using mobile phones in elections in developing countries: Opportunities and challenges. *Computing in research and development in Africa* (pp. 251–265). Springer.

Bajpai, N., & Biberman, J. (2020). *The COVID-19 economy: Building a sustainable future in the shadow of a crisis*. https://academiccommons.columbia.edu/

Bala, M., & Verma, D. (2018). Governance to good governance through e-governance: A critical review of concept, model, initiatives & challenges in India. *International Journal of Management, IT and Engineering, 8*(10), 244–269.

Banerjee, A., Duflo, E., Imbert, C., Mathew, S., & Pande, R. (2020). E-governance, accountability, and leakage in public programs: Experimental evidence from a financial management reform in India. *American Economic Journal: Applied Economics, 12*(4), 39–72.

Baroi, H. S., & Alam, S. (2021). Operationalizing the right to information act through e-governance in Bangladesh: Challenges and opportunities. *International Journal of Public Administration, 44*(8), 685–698.

Batta, M., Sethi, A., & Kaur, R. (2012). E-governance in e-administration. *International Journal of Computing & Business Research, 19*(1), 211–229.

Bhanti, P., Kaushal, U., & Pandey, A. (2011). E-governance in higher education: Concept and role of data warehousing techniques. *International Journal of Computer Applications, 18*(1), 875–895.

Bruegger, B. P., Hühnlein, D., & Kreutzer, M. (2007). Towards global eID-interoperability. *BIOSIG 2007: Biometrics and electronic signatures* (Vol. 108, pp. 127–140). Darmstadt, Germany.

Buchholz, B. A., DeHart, J., & Moorman, G. (2020). Digital citizenship during a global pandemic: Moving beyond digital literacy. *Journal of Adolescent & Adult Literacy, 64*(1), 11–17.

Chaudhary, A. (2020). E-governance for tribal India: Filling the communication gap. *International Journal of Social and Humanistic Computing, 3*(3–4), 308–316.

Chilena, M. (2013). Biblioteca Nacional de Chile. *Recuperado de*. http://www.memoriachilena.cl/602/w3-article-3320.html. Consulta 26 May 2016.

Chua, B. L., Al-Ansi, A., Lee, M. J., & Han, H. (2020). Tourists' outbound travel behavior in the aftermath of the COVID-19: Role of corporate social responsibility, response effort, and health prevention. *Journal of Sustainable Tourism, 29*(6), 879–906.

Cuijpers, C. M. K. C., & Schroers, J. (2014). *eIDAS as guideline for the development of a pan European eID framework in FutureID*. Gesellschaft für Informatik e.V. (GI).

Das, R. K., Patnaik, S., Padhy, A. K. & Mohini, C. (2014). Service oriented layered approach for e-governance system implementation. In *2014 International Conference on Information Technology* (pp. 293–298). IEEE.

Education Argentina. (2022). *Portal*. https://www.educ.ar/. Accessed 24 January 2022.

Elavarasan, R. M., & Pugazhendhi, R. (2020). Restructured society and environment: A review on potential technological strategies to control the COVID-19 pandemic. *Science of the Total Environment, 725*, 138858.

Garson, D. G. (2006). *Public information technology and e-governance*. Jones and Bartlett Publishers.

Giri, S., Shakya, S., & Pande, R. N. (2018). E-governance implementation: Challenges of effective service delivery in civil service of Nepal. *Global Journal of Computer Science and Technology, 18*(3), 1–9.

Gogo, E. O., Saidi, M., Ochieng, J. M., Martin, T., Baird, V., & Ngouajio, M. (2014). Microclimate modification and insect pest exclusion using agronet improve pod yield and quality of French bean. *HortScience, 49*(10), 1298–1304.

Goyal, S., Pillai, A., & Chauhan, S. (2021). E-governance using mobile applications: A case study of India during the COVID-19 pandemic. *Australasian Journal of Information Systems, 25*, 1–29.

Gupta, P. J., & Suri, P. (2017). Measuring public value of e-governance projects in India: Citizens' perspective. *Transforming Government: People, Process and Policy, 3*(1), 23–35.

Gupta, R., Pal, S. K., & Pandey, G. (2020). *A comprehensive analysis of COVID-19 outbreak situation in India*. MedRxiv. https://doi.org/10.1101/2020.04.08.20058347

Hooda, A., & Singla, M. L. (2020). Core–competencies–a key to future–oriented and sustainable e-governance implementation: A mixed method research. *Transforming Government: People, Process and Policy, 15*(1), 80–107.

Janssen, M., & Van Der Voort, H. (2020). Agile and adaptive governance in crisis response: Lessons from the COVID-19 pandemic. *International Journal of Information Management, 55*, 102180.

Karmakar, J. (2020). Understanding the role of e-governance in urban areas of west Bengal. In *Role of IT-ITES in economic development of Asia* (pp. 95–108). Springer.

Khanna, A. (2020). Impact of migration of labour force due to global COVID-19 pandemic with reference to India. *Journal of Health Management, 22*(2), 181–191.

Korea Times. (2021, December 11). *Travelers upset by 'sudden' announcement about 10-day mandatory quarantine*. https://www.koreatimes.co.kr/www/nation/2021/12/119_319926.html

Kuzma, J. (2010). Asian government usage of web 2.0 social media. *European Journal of ePractice, 9*, 1–13.

Lee-Geiller, S., & Lee, T. D. (2019). Using government websites to enhance democratic e-governance: A conceptual model for evaluation. *Government Information Quarterly, 36*(2), 208–225.

Malik, B. H., Shuqin, C., Mastoi, A. G., Gul, N., & Gul, H. (2016). Evaluating citizen e-satisfaction from e-government services: A case of Pakistan. *European Scientific Journal, 12*(5), 346.

Manoharan, A. P., Ingrams, A., Kang, D., & Zhao, H. (2021). Globalization and worldwide best practices in e-government. *International Journal of Public Administration, 44*(6), 465–476.

Melitski, J. (2019). Digital governance services. *E-government and Information Technology Management: Concepts and Best Practices, 1*(1), 83.

Mittal, P., & Kaur, A. (2013). E-governance: A challenge for India. *International Journal of Advanced Research in Computer Engineering & Technology, 2*(3), 56–67.

Mohammad Ebrahimzadeh Sepasgozar, F., Ramzani, U., Ebrahimzadeh, S., Sargolzae, S., & Sepasgozar, S. (2020). Technology acceptance in e-governance: A case of a finance organization. *Journal of Risk and Financial Management, 13*(7), 138.

Mukherjee, K., & Sahoo, G. (2010). Cloud computing: Future framework for e-governance. *International Journal of Computer Applications, 7*(7), 31–34.

Mutlugün, M., & Adalier, O. (2009, October). Turkish national electronic identity card. In *Proceedings of the 2nd International Conference on Security of Information and Networks* (pp. 14–18). National Center for Biotechnology Information.

New Zealand Herald. (2022, January 22). *COVID 19 Omicron: NZ moves to red light tonight, PM Jacinda Ardern confirms nine cases with variant*. https://www.nzherald.co.nz/nz/covid-19-omicron-nz-moves-to-red-light-tonight-pm-jacinda-ardern-confirms-nine-cases-with-variant/FPYXEZCL3IS2R3I5RPGAI4IFOA/

Noack, T., & Kubicek, H. (2010). The introduction of online authentication as part of the new electronic national identity card in Germany. *Identity in the Information Society, 3*(1), 87–110.

Pawar, N., Ghodke, K., Raman, P., Kotangale, S., Gupta, R., & Shelke, N. (2016). A critical study of implication of render billing in e-governance services. *IOSR Journal of Computer Engineering, 1*(1), 56–61.

Rahman, H. (2010). Framework of e-governance at the local government level. *Comparative e-Government, 37*(1), 23–47.

Ray, S., & Mukherjee, A. (2007). Development of a framework towards successful implementation of e-governance initiatives in health sector in India. *International Journal of Health Care Quality Assurance, 6*(2), 35–44.

Royal Oman Police. (2022). *Government website*. https://www.rop.gov.om/english/. Accessed 24 January 2022.

Shahi, G. K., Dirkson, A., & Majchrzak, T. A. (2021). An exploratory study of Covid-19 misinformation on twitter. *Online Social Networks and Media, 22*, 100104.

Shukla, P. S., & Mathur, M. (2020). Conceptualizing the role of data analytics and technology in e-governance: An insight. *International Journal of Business Analytics (IJBAN), 7*(2), 1–12.

Simpson, A., & Park, S. (2013). The Asian Development Bank as a global risk regulator in Myanmar. *Third World Quarterly, 34*(10), 1858–1871.

Suárez, G., Espinosa-Aranda, J. M., Cuéllar, A., Ibarrola, G., García, A., Zavala, M., Maldonado, S., & Islas, R. (2018). A dedicated seismic early warning network: The Mexican Seismic Alert System (SASMEX). *Seismological Research Letters, 89*(2A), 382–391.

The Hindu. (2020, June 5). *Coronavirus lockdown | 26 lakh migrant workers in halfway houses, says official data.* https://www.thehindu.com/news/national/coronavirus-lockdown-26-lakh-migrant-workers-in-halfway-houses-says-official-data/article31751222.ece

Tikhomirova, O. (2020). E-governance and social inclusion of entrepreneurship and businesses: Toward the social inclusive digital society. *International Journal of E-Entrepreneurship and Innovation (IJEEI), 10*(2), 1–25.

Ullah, A., Pinglu, C., Ullah, S., Abbas, H. S. M., & Khan, S. (2021). The role of e-governance in combating COVID-19 and promoting sustainable development: A comparative study of China and Pakistan. *Chinese Political Science Review, 6*(1), 86–118.

UN E-Governance Report. (2020). *Digital government in the decade of action for sustainable development.* https://www.un.org/development/desa/en/news/administration/2020-egovernment-survey.html

Yadav, N., & Singh, V. B. (2013). E-governance: Past, present and future in India. *International Journal of Computer Applications, 53*(7), 36–48.

PART III

New Normal Policies for the Digital Transformation of Enterprises

CHAPTER 7

Role of Enterprises and Their Digital Transformation in Post-pandemic Sustainable Development

Vasiliki Vrana and Subhankar Das

Introduction

The technological revolution was already taking place before the global pandemic of COVID-19. It has accelerated the pace of change in various sectors of the economy and society (Mondal et al., 2022a). This document aims to analyze the potential of digital transformation to promote the reconstruction and reactivation of the regions affected

V. Vrana
Department of Business Administration, School of Economics and Administration, The Campus of Serres, International Hellenic University, Serres, Greece
e-mail: vrana@ihu.gr

S. Das (✉)
The Honors Programme, Southern Star Institute of Management, Duy Tan University, Da Nang, Vietnam
e-mail: subhankardas@duytan.edu.vn

S. R. Mondal et al. (eds.), *New Normal in Digital Enterprises*,
https://doi.org/10.1007/978-981-19-8618-5_7

by the pandemic (Tiwari & Mondal, 2022). It also explores the various conditions dragging the area since the pandemic started. Digital transformation can help countries in the region overcome the COVID-19 pandemic and develop new development opportunities (Mondal & Das, 2021). It can provide them with new tools and resources to address various issues, such as improving their public institutions' efficiency and effectiveness and production systems. Before the pandemic started, the use of digital technology had already accelerated (Sharma & Das, 2021). The rapid spread of broadband internet and mobile phones enabled many individuals to connect to the digital world. Although many companies in the region have gained access to digital technologies, they are not yet taking advantage of the full potential of these tools (Das, 2021a). The productivity gains from digital transformation are not reflected in the country's economic performance. Despite the positive effects of the pandemic, the widespread use of digital technologies is still not widely spread worldwide (Siri & Das, 2021). Various factors, such as the lack of regulatory frameworks and good training programs, are still needed to enable a profound digital transformation (Duman & Das, 2021; Ravi & Mondal, 2021). Unfortunately, the COVID-19 pandemic has affected the world negatively.

This chapter aims to discuss the various advantages of digital transformation and cloud computing in helping countries in the region recover from the pandemic. It argues that these technologies can help them improve their productivity and address their development gaps. Many reports and analysts had urged policymakers to be more proactive in addressing digital gaps before the COVID-19 pandemic. However, despite the urgency, the progress has not been as fast as expected. Despite the adverse effects of the pandemic, the increasing awareness about the importance of digital transformation is helping countries in the region move toward recovery (Yegen & Mondal, 2021). The pandemic has also made people more aware of the importance of technology in helping them improve their lives. It has also highlighted the various opportunities digital transformation can provide for governments and individuals (Das, 2021b). Despite the various shortcomings of digital transformation, it is still a crucial factor that can help countries improve their resilience and build a more competitive and prosperous society (Das, 2020a). The rapid emergence and evolution of COVID-19 have created new opportunities for countries in the region to address their digital gaps and improve their resilience (Sharma et al., 2020). The years 2022–2025 will be a crucial time for policymakers and the private sector to step up their efforts in addressing these issues (Mondal, 2020a).

Digital Transformation and Sustainability

It's time for businesses to start taking them seriously of their environmental responsibilities. According to various researchers, more than 50% of consumers want sustainable brands (Das, 2021c; Duy et al., 2020; Mondal, 2021). Unfortunately, despite the increasing number of consumers demanding sustainable products, only 35% of businesses have taken the necessary steps to implement their environmental strategies (Singh et al., 2020; Siri et al., 2020; Van et al., 2020). Through digital transformation, businesses can also benefit from the implementation of sustainable practices. Different studies revealed that by integrating their digital transformation efforts, some of the most transformative companies were able to create such a change (Das, 2020b; Mondal, 2020b; Sharma & Das, 2020). Despite the number of initiatives being made by companies to improve their sustainability reporting, only half of them is willing to make the necessary changes in their operations (Das & Mondal, 2016). These companies were able to achieve sustainable revenue growth by implementing their digital transformation initiatives. From 2018 to 2021, these firms saw an estimated annual revenue growth of over 50% (Nadanyiova & Das, 2020). Despite the increasing number of regulations and consumer expectations regarding sustainability, only 39% of companies are currently developing more energy-efficient and sustainable products and services (Behera et al., 2019; Gupta et al., 2019; Mohanty et al., 2019). This is a clear indication that businesses need to start taking the necessary steps to improve their environmental performance (Jain et al., 2018; Mondal et al., 2017; Singh & Das, 2018).

Despite the lack of specific solutions, many companies still believe that they have the technological capabilities to achieve sustainable operations. According to executives, implementing artificial intelligence, the Internet of Things, and mobile technologies are some of the key factors that businesses need to consider to achieve their goals. These companies are known as the transformative trailblazers because they use sustainability as a competitive advantage by implementing digital transformation.

Transformational leaders are focused on driving change and innovation through their company's sustainable operations. They are more likely to adopt new technologies like blockchain and artificial intelligence.

The chief information officer (CIO) is also responsible for leading the implementation of this strategy. They are also involved in the company's corporate governance and digital transformation. Transformational leaders also work closely with their supply chains and ecosystems to achieve their goals.

Despite the challenges that sustainability can present, there is a path forward. One of the most critical factors in achieving sustainable success is listening to both the employees and customers. This is also a component of the transformational journey. According to the authors, businesses that engage their employees and customers are more likely to get the input they need to shape their sustainability agenda (Cao & Posen, 2022). This allows them to identify new opportunities and develop practical solutions. Leaders committed to continuous improvement can harness the power of their customers, and employees can become successful in their industries (Freixanet & Federo, 2022).

The coronavirus pandemic has created a new world where the status quo no longer holds. It has caused widespread suffering and has forced us to rethink how we live. According to the UNDP, global human development could decline for the first time in two decades. The world has been through several crises over the years, such as the financial crisis of 2007–2009. However, development gains have been recorded year on year despite the hard times. The coronavirus pandemic has created an enormous opportunity for the global community to reach the goals of the Sustainable Development Agenda (George et al., 2021). These include ensuring a stable and prosperous life for all people. Unfortunately, the coronavirus pandemic has brought a grievous blow to the global community. It also shows us the need to step up and address the various challenges we face. According to the findings of several studies, most developing countries are already dealing with the effects of the pandemic.

The number of undernourished people has significantly decreased over the years. However, despite the progress made, there are still over 800 million people who are severely undernourished (Hafner & Pidun, 2022). The COVID-19 pandemic has exposed the weaknesses of the global food supply chains. Despite the progress that China and India have made in reducing poverty, over 700 million people still live on less than US$1.90 a day (Jahanbakht & Mostafa, 2022). According to an estimate by Oxfam, the crisis could push over a billion people back into poverty.

The UNDP is leading the response to COVID-19's socio-economic effects. Around half of the global workforce works in the informal economy. According to the ILO, almost one in six young people has lost their jobs since the pandemic. According to the UNDP, over a billion students are affected by school lockdowns (Jahanbakht et al., 2021). The pandemic has also re-emphasized ensuring that all children have access to education. Due to the nature of the elections, many countries have postponed their referendums and national elections. It could lead to an increase in the risk of unrest.

The rise of the digital and e-commerce sectors was triggered by COVID-19. In 2020, we will look back and see how COVID-19 changed everything. COVID-19 has accelerated digital transformation and boosted e-commerce's share of global retail trade (Leppänen et al., 2021). As consumers and businesses became more digital, they purchased more goods and services online. A new report released by eTrade and UNCTAD highlights how COVID-19 has accelerated the global e-commerce industry's growth (Parast, 2017). However, she noted that the digital transition would have lasting impacts on society and daily lives. Developing countries should also play a role in the digital economy.

Although governments have focused on short-term measures to address the pandemic, they have also started to consider long-term strategies to support recovery (Song, 2022). For instance, in the Caribbean and Latin America, governments have begun to protect individuals and businesses. In Africa, Senegal launched an awareness campaign on the importance of e-commerce. In Asia, Indonesia launched a program to accelerate the digital transformation of small and medium enterprises.

The rise of cloud computing has become an essential component of any enterprise's digital transformation strategy. It is also a good indicator of how tech buying is performing overall. In the first quarter of 2020, global spending on cloud infrastructure services reached \$31.0 billion (Yoruk & Jones, 2020). The rise in spending was driven by the COVID-19 pandemic and the need for faster access to computing resources. Many organizations have turned to public cloud services for their burst capabilities. These services allow them to respond to a spike in demand quickly.

Risk of Digital Transformation and Investment

However, despite the strong growth, the research also revealed that many enterprises are still cautious about their digital investment plans (Das et al., 2022a). According to researchers, many cloud users switch to three-month contracts to lower their costs and keep their employees working during state lockouts (Das et al., 2022b). This shift in usage patterns has caused businesses to put digital transformation projects on hold (Mondal et al., 2022b). It has caused many organizations to rethink their business model. However, it is still not clear if this will lead to a complete change in how they do business. Many of them still believe that their core business model is geared toward their customers.

Despite the changes in consumer behavior, there are still plenty of potential consumers out there. According to various studies, people no longer buy CDs anymore. At this moment, people are still conducting business meetings using Zoom instead of face-to-face meetings (Ngoc-Vinh et al., 2022). It has caused many business owners to start thinking about reaching their existing customers. As businesses evolve their operations, they should also consider different acquisition channels and develop new business strategies (Tien-Dung et al., 2022a). They should also be open to new ideas and technology. Even the smallest businesses can benefit from digital transformation. For instance, if a restaurant serves food and drinks, it can use the various platforms offered by Door Dash, Yelp, and other online businesses to gain a competitive advantage. Most companies have at least one operational system: email, CRM, and accounting. They can also use the various integration tools available on the market to automate specific processes (Minh-Nhat et al., 2022a). According to researchers, not implementing digital transformation is the most significant risk for small business enterprises.

Being Ready for Digitalization; a Much-Needed Mindset

The biggest threat to a business's success is not a lack of digital strategy but a lack of readiness for change. Therefore, it is so important to have a current and effective system. According to various studies, customers will leave a company if it doesn't adapt quickly to the pandemic's changes (Nguyen et al., 2022). The current situation has accelerated the pace of change, which means that many companies will not be able to implement their digital strategies after the pandemic ends (Minh-Nhat et al., 2022b). One of the most critical factors that businesses need to consider when developing a digital strategy is managing risk. Doing so will allow them to keep delivering on their promises while keeping their customers satisfied.

Due to the rapid pace of change, many companies will not be able to implement their digital strategies after the pandemic ends. Various research also stated that customers would leave a company if it didn't adapt quickly to the pandemic's changes (Tien-Dung et al., 2022b). One of the most critical factors that businesses need to consider when developing a digital strategy is managing risk. Doing so will allow them to keep delivering on their promises while keeping their customers satisfied.

Globalization vs. Localization

The various characteristics of the multinational enterprise (MNE) make it worthy of attention. These include its expansive organizational structure and the different competing interests it must address. The various factors contributing to the rise of digital disruption and pandemics are also challenging the corporate advantage of MNEs. These factors are driving them to become more global and local.

In business, globalization refers to distributing information and products across borders. On the other hand, localization is a concept focused on a specific target audience. Getting global is a rewarding experience, especially if you're an entrepreneur. The rise of the internet has made it possible for people to connect and reach customers globally. However, according to a study, over 40% of internet users won't buy from websites that aren't in their native language. Another study revealed that over half of Google searches are conducted in languages other than English. Aside from establishing a global brand, globalization also involves developing a strategy that focuses on localization. The concept of globalization is a process that consists of the spread of knowledge and ideas across a global scale. It's a process that can be described as free trade. In business, companies can easily access resources and ideas from other countries. Globalization aims to allow companies to access new ideas and technologies from other countries. However, cultural and political differences can prevent them from fully realizing the potential of this process. It can be very challenging for businesses trying to enter foreign markets.

The concept of globalization is a powerful driver for business growth. It calls for increased interaction and integration among nations, which can help create a more balanced and competitive environment. Various elements go into making a successful globalization strategy. Due to the rise of economic and cultural globalization, countries have become more politically connected. It has resulted in them working together to create trade agreements and allow the movement of capital. Aside from being beneficial for business, globalization can also have environmental and economic implications. Doing business in other countries can harm the environment. The concept of globalization enables companies to access a wider variety of resources, such as skilled labor and materials. It can also help them improve their operations and supply chains. The concept of globalization and business ethics are rooted in the idea of diversity. Having a framework that includes ethical views and practices can help companies establish a more inclusive culture.

Getting the most out of your online presence is essential, but it's also about addressing the cultural and language barriers that keep people from interacting with your company. While localization is focused on translating content, it also involves addressing the various cultural and language barriers that keep global customers from interacting with your company. One of the most effective ways to do this is through website localization. Although localization is only one component of the globalization puzzle, it also plays a massive role in developing a successful strategy. Having the proper mix of localization elements in your business operations can help ensure that your customers can seamlessly connect with you. For global businesses, localization is a vital component of their strategy as it allows them to adapt their content to the local market. This process involves translating a text into another language while preserving the local culture's meaning. Aside from translating content, localization can also include adapting various website elements to the local market. It can consist of colors, designs, and payment methods.

The localization elements are those that a company needs to address when developing content for a global audience. It includes cultural factors such as language and style. The data and transactional information that customers need to understand to purchase a product should be included in localization. It ensures that they have the necessary information to make an informed purchase. The cultural elements of localization help customers understand the market and location of their target audience. Enterprises can use localized content in various ways, such as email campaigns and social media posts. Both the concept of globalization and localization are related practices that help companies establish a long-term strategy for reaching their global audience. Localization is a vital part of this process, as it shows a commitment to the audience.

Being able to address your customers' needs in a way that they are comfortable with is very important for a company. Having the necessary localization tools can help improve the customer journey and make them more likely to purchase. While developing a strategy to go global, it's also essential to include localization to address the various cultural and language barriers encountered. On the other hand, when developing content, localization is only one component of a comprehensive strategy. Researchers understand the importance of both localization and globalization, and we can help you navigate the various steps involved in implementing both concepts.

Blending Opportunities of Globalization, Internationalization, Localization, and Transformation for Enterprises (GILT Effect)

Getting started with internationalization and globalization is a vital step in any company's journey to expand globally. Understanding the differences between these two actions is a must to make informed decisions when taking your company to the next level. One of the most important ways to distinguish these two concepts is to visualize them as rings that gradually get smaller and more focused. At the bottom of the ring is the big-picture step of globalization, which involves the procedures and policies that your company will adopt to expand globally. The second step is internationalization, which consists of the design and development of products and services geared toward expanding into international markets. The concepts of internationalization and globalization are part of the same structure, but they are built on one another to achieve the company's goals. In terms of data and technology, globalization refers to preparing products and services that are designed to be more widely used and perceived by different audiences. This process involves opening international borders and promoting the flow of goods and services.

A successful globalization process involves a wide range of factors that affect the operation of a company. Some of these include the rules that you must follow to operate in other countries, the type of products and services you can offer, and the branding and marketing of your company. Online marketplaces like Amazon and Netflix allow users to shop for products and services from different regions. Both companies offer different types of content and services in other countries. The goal of globalization is to make sure that the products and services you offer are adapted to the needs of different audiences. Developing and implementing an internationalization strategy is a process that helps organizations create products and services that can be easily translated and localized for other markets. This step is usually performed in stages to ensure that the finished product can be adapted to the target audiences. Before implementing an internationalization strategy, the design and development stages must be completed to ensure that the finished product can be easily translated and localized. For instance, removing the text from the Starbucks logo made it easier for people to recognize its brand around the world easily.

In 1991, Kentucky Fried Chicken, a fast-food chain known for its chicken, became KFC. To protect its brand, the state of Kentucky required any company that uses the word "Kentucky" for commercial purposes to pay a licensing fee. KFC then introduced a complete brand overhaul, including new packaging and image. The company noted that the name change made it easier to say "KFC" worldwide. It also helped in taking the focus off fried chicken and focusing on other menu items. The success of fast-food chains such as McDonald's and KFC can be attributed to their consistent brand image and familiarity with what they serve. Thinking beyond the country and language of your company's original brand is also essential to ensure that the finished product can be easily translated and localized. Enterprise's various parts are often translated into a different language and cultural standards to meet customers' needs in each region or country. This process can be done through website content, mobile apps, and branding.

Besides digital transformation, localization is also a process that aims to give your company a look and feel representative of the people who live in the region where it's located. Aside from translating, localization also ensures that the content and services are presented most appropriately. This process can include using local address formats and displaying the correct units of measurement. One of the most successful localization projects is Airbnb, an online marketplace for lodging. They define it as the deliberate creation of products and services that are globally consistent.

Airbnb was founded on the idea that people should be able to spend their nights in private homes around the world. To fulfill this mission, Airbnb has to be able to provide its guests and hosts with the same experiences in their local languages. It also aims to respect the local expectations and cultural norms. Over the years, Airbnb has expanded its global reach by translating its website into over 60 different languages. It ensures that the company's content is always up-to-date and relevant to the needs of its customers. If both the product and the company's overall operations are highly localized, it helps achieve long-term success. For instance, if you're planning on visiting Japan, the experience of using Airbnb should be the same in Japan as it is in Canada. McDonald's is an example of a company that has successfully implemented localization strategies. Through their efforts, they have introduced new products and menus in different regions. They also regularly test new concepts to keep up with the changing cultural norms. Artificial intelligence (AI)-powered virtual assistants such as Amazon Alexa, Google Assistant, and a large portion of the population commonly uses Siri. These tools collect vast amounts of data to provide the best possible service to their users. One of the main goals of internationalization is to enable the assistants to recognize different languages. They can also switch between them by coding their systems to allow them to do so. Figure 7.1 represents the effect of globalization, internalization, and localization.

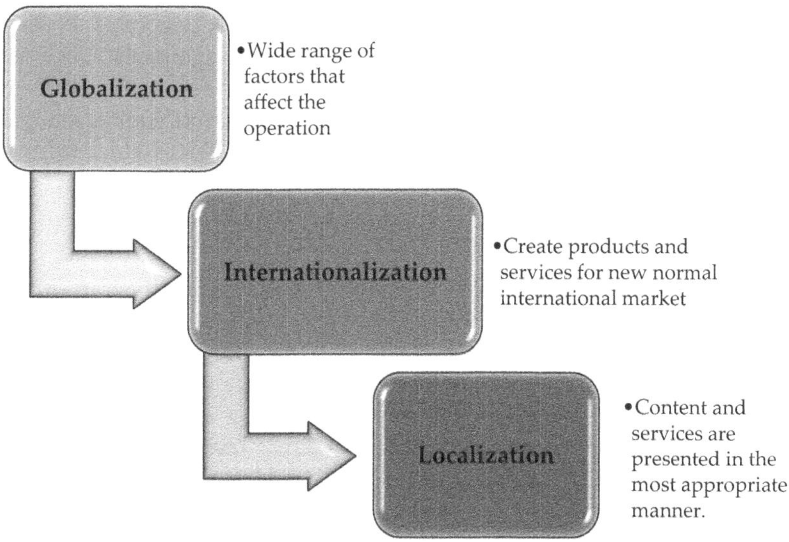

Fig. 7.1 GIL concept and how it brings transformation (*Source* Authors' conception)

GILT Effects on Digital Enterprises

1. Easy penetration to the post-pandemic new regular market: Getting the proper planning for the three phases of your business will help you reach your goals. There are several reasons why planning for these steps is essential. Before you start thinking about internationalization, consider the market from the ground up. Doing so will allow you to take advantage of the opportunities in new markets and avoid cultural mistakes. Having a global strategy will also help you avoid costly mistakes.
2. Customer-friendly operation: imagine having a global perspective is very important for any company, as it can help them reach a wider audience and increase their sales. Developing and maintaining effective international marketing content is also very important, as it can help boost your sales and increase your chances of success. Unfortunately, many businesses fail to develop and maintain effective international websites and apps.
3. Website/application for profitable ways of customer dealing: Having a well-designed and user-friendly website and app will help increase your conversion rates and attract more potential customers. Having a positive customer experience will also encourage repeat business. A localization service will also help you establish yourself in different regions and countries. It will allow you to expand your reach and reach more people.

Market Model and Value Creation

The pandemic brought people together to realize that they are interconnected, which led to increased awareness about the risks of climate change. According to a survey conducted by the Boston Consulting Group, almost hundred percent of respondents believe that their actions can help reduce greenhouse gas emissions and protect biodiversity. In a study conducted in the US, UK, and China, researchers found that almost all consumers expect companies to reduce their environmental footprint. They also hope to contribute to positive impact projects such as reforestation and ocean plastic clean-ups. Due to the increasing number of investors demanding transparency and environmental activism, many companies have started to rethink their financial practices. It has led to new incentives for companies to reduce their environmental impact.

Climate justice and consumer awareness have also risen due to increasing calls for corporate social responsibility. One of the most common criticisms about how companies do business is how they extract revenue from their local communities.

Through the digitization of corporate behavior, the public has been able to gain a deeper understanding of how companies treat their various stakeholder groups. As a result, many companies have started to implement comprehensive environmental monitoring and reporting systems. Getting the necessary information about the impact of their activities is very important to meet the targets set by the government. Digitization has also helped companies lower their costs by allowing them to monitor and analyze their data in real time. The challenge for large-cap companies is that their traditional models of creating appropriate value are being challenged by the increasing number of calls for global solutions. It is also due to the growing awareness of the climate crisis. The ever-increasing scope and depth of environmental monitoring and reporting have allowed large companies to address their concerns. However, the associated costs are still a significant challenge for them. Due to the various factors that affect the operations of large companies, such as political tensions, market conditions, and technological changes, they are also facing increasing challenges.

The pandemic brought people together to realize that they are interconnected. It also raised awareness about the risks of climate change and the need to reduce waste. According to a survey conducted by the Boston Consulting Group, almost 90% of respondents believe that their actions can help reduce greenhouse gas emissions. They also believe that the pandemic will only serve as a learning tool for investors as climate change is expected to become more severe. In a study conducted in 2011, researchers found that almost 80% of consumers in the US, UK, and China expect companies to reduce their greenhouse gas emissions by adopting sustainable practices.

Due to the increasing awareness about the risks of climate change, investors are demanding more transparency from companies. They are more willing to go after executives who fail to take climate risk seriously. It has prompted many companies to rethink their financial practices. Climate justice advocates cite the increasing nationalism as one factor that has encouraged them to call for more transparency from MNEs. They also believe that the company's extractive structure undermines the relationship between the local community and its headquarters.

The goal of achieving a sustainable environment is to collect timely and credible information about the impact of their activities. Due to the increasing visibility of corporate behavior, companies are now required to put in place comprehensive environmental monitoring and impact management systems. This is expected to fuel the development of a digital industry that will allow them to report on their environmental and social governance (ESG) activities. Using digital technologies, companies can reduce the cost of monitoring their ecological activities. The challenge for the global non-profit sector is that the standard models of how multinational companies create and appropriate value is being challenged. It is due to the growing awareness of the climate crisis and the need for global solutions.

Due to the various factors that affect the operations of multinational companies, such as political tensions, market conditions, and technological changes, they are becoming more complex to manage. The increasing scope and depth of environmental monitoring and reporting is helping multinational companies address these concerns. However, the associated costs are still considered to be a significant disadvantage against their local competitors.

Streamlining the Operation in the Post-pandemic World

The pandemic has dramatically impacted the way people work. Using digital tools, employees have become more productive and have shifted the firm's focus from a physical space to a digital one. The shift to digital processes is not expected to be fully reversible. It can create micro-diffusions of power between different levels of authority, which raises theoretical questions about the role of central authority in the workplace. Since there are varying cultural and national preferences regarding working from home, it is unclear how MNEs can effectively implement the new digital processes. To maintain its competitive advantage, it will need to improve its governance practices. Climate risk and the pandemic have highlighted the vulnerability of global supply chains to various factors, such as natural disasters and political changes. It has reignited the debate about the use of digital technologies to improve the efficiency and resilience of the supply chain.

The combination of climate and digitization trends has created new theoretical challenges related to coordinating and collaborating for social, organizational, and network resilience. One of these is blockchains, which can reduce the need for monitoring inter-organizational relationships. Due to the availability of modern equipment and the reduction of coordination costs, economies of scale have become more resilient. The use of digital tools has made it possible for people to access and use productive equipment in the cloud. The rise of digital technology has significantly altered the basis of competition. It has also created new challenges for large organizations, typically less agile and has traditional routines and practices.

While large organizations have a competitive advantage due to their economies of scale, digital companies are smaller and have a stronger advantage using open resources. In contrast, digital platforms such as Airbnb and Uber have significantly reduced the number of people they hire while increasing their valuations. The operating leverage of shared resources has raised various ethical and environmental concerns. For instance, it could affect the resource footprints of nations and communities.

Explore the Opportunities Created by Digital Disruption

Due to the increasing number of environmental and health crises, the demand for sustainable and ESG solutions has become more prevalent among investors and customers. It has prompted more companies to explore new approaches to address these issues. Despite the importance of sustainability, it remains a weak link for delivering superior value. Instead of reducing negative externalities, companies should instead consider adopting more effective and cost-efficient measures. The solution to the growing environmental and health crises may need more effective and sustainable regeneration. According to a report released by UNEP, both public and private sectors need to increase their investments in nature by 2030 and 2050, respectively. The private sector is expected to only invest about 18 billion dollars annually in regeneration, while the public sector is expected to spend around 1.5 trillion dollars on digital transformation. This report highlights the importance of digital and sustainability convergence in addressing the issues of environmental and health crises. It also shows that the scale of the investment needed for regeneration can be met by rebuilding natural capital. Although there is a lot of conceptual work on sustainable digital transformation, there is also a lack of empirical evidence on the effectiveness of novel business models and organizational designs.

The market for nature-based solutions is expected to increase due to the increasing number of credible and effective actions companies take. Due to their immense scale and impact, many companies are expected to go beyond their zero waste and carbon neutrality initiatives and start developing value-based strategies. This is because they can identify and implement effective and sustainable actions that can directly and positively impact the world. One of the most effective ways to implement value-based strategies is by aligning the concept of regeneration with the company's strategic growth goals. This can help service companies reduce their energy and manufacturing-related externalities. It can be done by developing effective digital monitoring and reporting tools.

The increasing number of cross-disciplinary applications that are making the world a better place has raised new questions about the concept of value capture and creation. Various theoretical perspectives can be used to analyze the various aspects of corporate purpose and sustainability. These include stakeholder and shareholder perspectives, fairness theories, procedural justice, and diversity. Using digital tools and processes, MNEs can create a sense of collective ownership by incorporating positive impact into their existing operations. For instance, by implementing digital automations, they can protect natural reserves and restore forests. As sustainability becomes a competitive differentiator for many companies, they can start growing their handprint by developing regenerative business models that can create public value while also contributing to the company's bottom line.

Conclusion

Using digital tools and services, MNEs can create a stronger connection with their customers and communities by incorporating their goodwill and reputation into their transactions. For instance, automating specific processes can help protect natural resources and improve the environment. The emergence of new applications that can transform the world has raised further questions about how we look at value creation and value capture. As sustainability becomes a competitive differentiator for MNEs, they can start developing new business models that are circular and regenerative. These models can help them create public value while also contributing to developing their employee and customer loyalty.

Due to the increasing importance of climate and pandemic issues in the corporate world, the convergence of sustainability and digital technologies has created new opportunities for MNEs to impact the planet positively. These new tools can help them make a lasting impact on the earth. It provides an opportunity for management scholars to study the various factors that influence sustainable development and how they can be applied to the operations of multinational companies. These findings could then be used to develop new theoretical frameworks and approaches to the design and operation of global firms.

References

Behera, M., Dash, M., Ayasa Kanta Mohanty, A. K., & Das, S. (2019). Entrepreneurial ecosystem: A geographical upliftment. *Revista Espacios, 40*, 26–36. http://www.revistaespacios.com/a19v40n01/19400126.html

Cao, Z., & Posen, H. E. (2022). When does the pre-entry experience of new entrants improve their performance? A meta-analytical investigation of critical moderators. *Organization Science*. https://doi.org/10.1287/orsc.2022.1589

Das, S. (2020a). *Media impact of advertising on consumer buying behaviour a comparative study of different media*. Shiksha o Anusandhan University. http://hdl.handle.net/10603/273691

Das, S. (2020b). Innovations in digital banking service brand equity and millennial consumerism. *Digital transformation and innovative services for business and learning* (pp. 62–79). IGI Global. https://doi.org/10.4018/978-1-7998-5175-2.ch004

Das, S. (2021a). *Digital entertainment: The next evolution in service sector*. Springer Nature. https://doi.org/10.1007/978-981-15-9724-4.PrintISBN 978-981-15-9723-7.OnlineISBN978-981-15-9724-4

Das, S. (2021b). *Search engine optimization and marketing: A recipe for success in digital marketing* (1st ed.). CRC Press. https://doi.org/10.1201/978042 9298509

Das, S. (2021c). A systematic study of integrated marketing communication and content management system for millennial consumers. In *Innovations in digital branding and content marketing* (pp. 91–112). IGI Global. https://doi.org/10.4018/978-1-7998-4420-4.ch005

Das, S., & Mondal, S. (2016). *Yours digitally: Patanjali*. Available at SSRN: https://ssrn.com/abstract=2737299

Das, S., Mondal, S. R., & Sandhu, K. (2022a). Music logos drive digital brands: an empirical analysis of consumers' perspective. *Journal of Strategic Marketing*, 1–16. https://doi.org/10.1080/0965254X.2022.2098526

Das, S., Mondal, S., Puri, V., & Vrana, V. (2022b). Structural review of relics tourism by text mining and machine learning. *Marketing*, *8*(2), 25–34. https://doi.org/10.5281/zenodo.7358349

Duman, Ü. O., & Das, S. (2021). The dynamic rise of digital brands' market Mavens in digital entertainment: A complete know how for curious customers. In *Digital entertainment* (pp. 147–172). Palgrave Macmillan. https://doi.org/10.1007/978-981-15-9724-4_8

Duy, N. T., Mondal, S. R., Van, N. T. T., Dzung, P. T., Minh, D. X. H., & Das, S. (2020). A study on the role of web 4.0 and 5.0 in the sustainable tourism ecosystem of Ho Chi Minh City, Vietnam. *Sustainability, 12*(17), 7140. https://doi.org/10.3390/su12177140

Freixanet, J., & Federo, R. (2022). The complex interplay of firm innovation, internationalization and learning capability in driving firm performance: A configurational analysis. *Journal of Strategy and Management*. https://doi.org/10.1108/jsma-11-2021-0220

George, G., Haas, M. R., McGahan, A. M., Schillebeeckx, S. J. D., & Tracey, P. (2021). Purpose in the for-profit firm: A review and framework for management research. *Journal of Management*. https://doi.org/10.1177/01492063211006450

Gupta, D. K., Jena, D., Samantaray, A. K., & Das, S. (2019). HRD climate in selected public sector banks in India. *Revista ESPACIOS, 40*(11), 14–20.

Hafner, C., & Pidun, U. (2022). Getting family firm diversification right: A configurational perspective on product and international diversification strategies. *Journal of Family Business Strategy, 13*(1), 100456. https://doi.org/10.1016/j.jfbs.2021.100456

Jahanbakht, M., & Mostafa, R. (2022). The emergence of GVCs for frontier markets: Insights from the African mobile telecommunications industry. *Africa Journal of Management, 8*(1), 59–82. https://doi.org/10.1080/23322373.2021.2001287

Jahanbakht, M., Mostafa, R., & Veloso, F. (2021). Pre-entry experience, postentry adaptations, and internationalization in the African mobile telecommunications industry. *Organization Science*, 1–22. https://doi.org/10.1287/orsc.2021.1470

Jain, S., Jain, V., & Das, S. (2018). Relationship analysis between emotional intelligence and service quality with special evidences from Indian banking sector. *Revista ESPACIOS, 39*(33).

Leppänen, P., George, G., & Alexy, O. (2021). When do novel business models lead to high firm performance? A configurational approach to value drivers, competitive strategy, and firm environment. *Academy of Management Journal*. https://doi.org/10.5465/amj.2020.0969

Minh-Nhat, H. O., Majerova, J., & Das, S. (2022a). Taking steps to help the phoenix rise from ashes: A roadmap for digital enterprises to develop new sustainable goals. In *Sustainable Development and Innovation of Digital Enterprises for Living with COVID-19* (pp. 35–51). Springer. https://doi.org/10.1007/978-981-19-2173-5_3

Minh-Nhat, H. O., Nguyen, H. L., & Mondal, S. R. (2022b). Digital transformation for new sustainable goals with human element for digital service enterprises: An exploration of factors. In *Sustainable Development and Innovation of Digital Enterprises for Living with COVID-19* (pp. 85–103). Springer. https://doi.org/10.1007/978-981-19-2173-5_6

Mohanty, P. C., Dash, M., Dash, M., & Das, S. (2019). A study on factors influencing training effectiveness. *Revista Espacios, 40*, 7–15. http://www.revistaespacios.com/a19v40n02/19400207.html

Mondal, S. R. (2020a). *Factors influencing store image loyalty and satisfaction from customer perspective an empirical study in retail sector*. Shiksha o Anusandhan University. http://hdl.handle.net/10603/273688

Mondal, S. R. (2020b). A systematic study for digital innovation in management education: An integrated approach towards problem-based learning in Vietnam. In *Digital innovations for customer engagement, management, and organizational improvement* (pp. 104–120). IGI Global. https://doi.org/10.4018/978-1-7998-5171-4.ch006

Mondal, S. R. (2021). A systematic study of new age consumer engagement and exploration for digital entertainment for over-the-top platforms in various digital media. In *Innovations in digital branding and content marketing* (pp. 113–133). IGI Global. https://doi.org/10.4018/978-1-7998-4420-4.ch006

Mondal, S. R., & Das, S. (2021). Examining diabetic subjects on their correlation with TTH and CAD: A statistical approach on exploratory results. In *Machine learning and the internet of medical things in healthcare* (pp. 153–177). Academic Press. https://doi.org/10.1016/B978-0-12-821229-5.00007-0

Mondal, S. R., Das, S., Musunuru, K., & Dash, M. (2017). Study on the factors affecting customer purchase activity in retail stores by confirmatory factor analysis. *Revista ESPACIOS, 38*(61).

Mondal, S. R., Virgilio, F. D., & Das, S. (2022a). *HR analytics and digital HR practices*. Springer. https://doi.org/10.1007/978-981-16-7099-2

Mondal, S. R., Majerova, J., & Das, S. (2022b). Sustainable Development and Innovation of Digital Enterprises for Living with COVID-19. Springer. https://doi.org/10.1007/978-981-19-2173-5

Nadanyiova, M., & Das, S. (2020). Millennials as a target segment of socially responsible communication within the business strategy. *Littera Scripta, 13*(1), 119–134. https://doi.org/10.36708/Littera_Scripta2020/1/8

Ngoc-Vinh, N., Tien-Dung, P., & Das, S. (2022). Digitization of Business; Need for Recombination During COVID-19. In *Sustainable Development and Innovation of Digital Enterprises for Living with COVID-19* (pp. 1–17). Springer. https://doi.org/10.1007/978-981-19-2173-5_1

Nguyen, T. D. M., Mondal, S. R., & Das, S. (2022). Digital entrepreneurial transformation (DET) powered by new normal sustainable developmental goals (n-SDGs): Elixir for growth of country's economy. In *Sustainable Development and Innovation of Digital Enterprises for Living with COVID-19* (pp. 69–84). Springer. https://doi.org/10.1007/978-981-19-2173-5_5

Parast, M. (2017). Linking operations strategy to competitive priorities and firm performance. *Academy of Management Proceedings, 2017*(1), 16316. https://doi.org/10.5465/ambpp.2017.336

Ravi, S., & Mondal, S. R. (2021). Digital entertainment based do it yourself content & advertisement as a factor of driving force for trust & customer patronage. In *Digital entertainment* (pp. 127–146). Palgrave Macmillan. https://doi.org/10.1007/978-981-15-9724-4_7

Sharma, E., & Das, S. (2020). Measuring impact of Indian ports on environment and effectiveness of remedial measures towards environmental pollution. *International Journal of Environment and Waste Management, 25*(3), 356–380. https://doi.org/10.1504/IJEWM.2019.10021787

Sharma, E., & Das, S. (2021). Integrated model for women empowerment in rural India. *Journal of International Development*, 1–18. https://doi.org/10.1002/jid.3539

Sharma, E., Nigam, N., & Das, S. (2020). Measuring gap in expected and perceived quality of ICT enabled customer services: A systematic study of top ten retailers of India. *International Journal of Applied Systemic Studies, 9*(2), 159–184. https://doi.org/10.1504/ijass.2020.113260

Singh, L. B., Mondal, S. R., & Das, S. (2020). Human resource practices & their observed significance for Indian SMEs. *Revista ESPACIOS, 41*(7). http://www.revistaespacios.com/a20v41n07/20410715.html

Singh, S., & Das, S. (2018). Impact of post-merger and acquisition activities on the financial performance of banks: A study of Indian private sector and public sector banks. *Revista Espacios Magazine, 39*(26), 25.

Siri, R., & Das, S. (2021). A study on processing of information storage & use of new age consumers in digital wellness sector through story telling & creating interest. In *Digital entertainment* (pp. 45–63). Palgrave Macmillan. https://doi.org/10.1007/978-981-15-9724-4_3

Siri, R., Mondal, S. R., & Das, S. (2020). Hydropower: A renewable energy resource for sustainability in terms of climate change and environmental protection. In *The handbook of environmental chemistry*. https://doi.org/10.1007/698_2020_635

Song, M. (2022). Market competition and regulatory compliance in public, non-profit, and for-profit organizations. *Public Management Review*, 1–21. https://doi.org/10.1080/14719037.2022.2055774

Tien-Dung, P. H. A. M., Majerova, J., & Das, S. (2022a). Phases of possible recovery of digital enterprises in new normal business for living with COVID-19 Times: Opportunities for a new era in sustainable development goals. In *Sustainable Development and Innovation of Digital Enterprises for Living with COVID-19* (pp. 19–33). Springer. https://doi.org/10.1007/978-981-19-2173-5_2

Tien-Dung, P., Nguyen, X. D., & Das, S. (2022b). Sustainable goal achievement by digital revolution during and after pandemic; How much one wins and losses: A bird's eye view for future planning. In *Sustainable Development and Innovation of Digital Enterprises for Living with COVID-19* (pp. 133–147). Springer. https://doi.org/10.1007/978-981-19-2173-5_8

Tiwari, M., & Mondal, S. R. (2022). Technological dynamism of workforce management for effective education service delivery during and after COVID-19. In *HR analytics and digital HR practices* (pp. 199–215). Palgrave Macmillan. https://doi.org/10.1007/978-981-16-7099-2_8

Van, N. T. T., Vrana, V., Duy, N. T., Minh, D. X. H., Dzung, P. T., Mondal, S. R., & Das, S. (2020). The role of human–machine interactive devices for post-COVID-19 innovative sustainable tourism in Ho Chi Minh City, Vietnam. *Sustainability*, *12*(22), 9523. https://doi.org/10.3390/su12229523

Yegen, C., & Mondal, S. R. (2021). Sharenting: A new paradigm of digital entertainment of new age parenting and social media. In *Digital entertainment* (pp. 213–231). Palgrave Macmillan. https://doi.org/10.1007/978-981-15-9724-4_11

Yoruk, E., & Jones, P. (2020). Firm-environment alignment of entrepreneurial opportunity exploitation in technology-based ventures: A configurational approach. *Journal of Small Business Management*, 1–47. https://doi.org/10.1080/00472778.2020.1800354

CHAPTER 8

Sustainable Development and Digitalization in the Supply Chain for Environmental Protection: Changes and Trends

Jana Majerova and *Subhankar Das*

Introduction

The rapid emergence of digital technologies has dramatically affected our lives. The rise of e-commerce has turned us into consumers who can shop for products from anywhere. As a result, the logistics industry is experiencing strong growth. Due to the rise of digital consumers, the expectations of logistics businesses have become more demanding (Mondal et al., 2022). For instance, 50% of trucks fail to deliver on

J. Majerova
Department of Economics and Management, AMBIS University, Prague, Czech Republic
e-mail: jana.majerova@ambis.cz

S. Das (✉)
The Honors Programme, Southern Star Institute of Management, Duy Tan University, Da Nang, Vietnam
e-mail: subhankardas@duytan.edu.vn

S. R. Mondal et al. (eds.), *New Normal in Digital Enterprises*,
https://doi.org/10.1007/978-981-19-8618-5_8

their return journey. Due to the increasing number of customers and the pressure on logistics companies to deliver goods within 24 hours, they are forced to improve their efficiency to maintain their margins. Despite the various challenges that the logistics industry faces, digital transformation is the only solution identified to address these issues (Tiwari & Mondal, 2022). Aside from being a significant contributor to greenhouse gas emissions, the logistics industry is also a substantial contributor to the global transportation industry's carbon footprint (Mondal & Das, 2021). In 2016, the sector was responsible for over 25% of the country's greenhouse gas emissions. The logistics industry's continuous use of petroleum-based fuels is one of the main factors contributing to the increasing greenhouse gas emissions (Sharma & Das, 2021). This issue has become a growing concern for society. Despite the logistics industry's challenges, digital transformation is still the most significant opportunity for its players (Das, 2021a). According to a study conducted by the World Economic Forum, it is expected to provide the industry with $1.5 trillion in added value by 2025.

Despite the slow pace of technological development in the logistics industry, it is still important to note that many technological wonders can be utilized in the sector (Siri & Das, 2021). Modern warehouses and airports can be easily spotted with self-driving trucks. Companies such as Volvo and Mercedes-Benz are currently developing digital trucks that can drive themselves. According to a report by PWC, the adoption of self-driving trucks will have a substantial economic impact on the industry.

Due to the development of self-driving trucks, the number of human drivers will gradually decrease. Eventually, drones will be able to replace human drivers in last-mile delivery services. According to a survey conducted in Germany, 75% of the country's logistics companies believe that most warehouses will adopt data goggles by the end of the decade. Despite the lack of regulations, drones are still considered the biggest threat to the transportation industry (Ravi & Mondal, 2021). Several companies have already started implementing blockchain technology in their operations to improve the customer experience (Duman & Das, 2021). The increasing amount of data collected by the Internet of Things and artificial intelligence is expected to transform the logistics industry. This technology will allow companies to improve their operations by analyzing and reporting on the data they collect.

New business models are emerging due to the increasing number of Internet platforms that enable small and medium-sized enterprises to sell their products and service globally (Yegen & Mondal, 2021). The various benefits of digital transformation in the logistics industry are numerous (Das, 2021b). It can help improve the efficiency of the entire value chain and enable companies to lower their costs. This technology can also help them become more responsive to the changing market conditions (Sharma et al., 2020). Big data analytics is considered the most competitive advantage that companies can gain from using this technology (Sharma et al., 2020). It allows them to make better decisions and improve their operations (Mondal, 2020a). The World Economic Forum stated that companies should improve their data collection and use it to launch new services such as last-mile delivery. It will allow them to make informed decisions and improve their operations. Despite the presence of some of the industry's early adopters, many logistics companies still lag in their digital transformation (Das, 2020a). According to a survey, only two percent of German firms are ready to use drones (Das, 2021c; Mondal, 2021).

Although the low level of adoption of digital transformation in the transportation industry is not caused by a lack of interest or knowledge about the technology, it is also due to various factors (Duy et al., 2020; Van et al., 2020). For instance, some companies have legitimate doubts about the feasibility of specific solutions (Siri et al., 2020). Aside from regulations, other factors such as the lack of trust and collaboration among various parties in the supply chain can also prevent companies from adopting digital transformation (Singh et al., 2020). For instance, in most cases, the exchange of data between different players in the supply chain can be made more secure by having a common standard (Das, 2020b; Sharma & Das, 2020). Despite the various advantages of digital transformation, many companies in the transportation industry still face significant risks due to their lack of preparedness (Mohanty et al., 2019; Mondal, 2020b; Nadanyiova & Das, 2020). They must adopt the necessary changes if they want to survive in the future.

Environmental Sustainability and Intelligent Logistics

The logistics industry is responsible for 13% of global greenhouse gas emissions. According to EPA in 2016, the emissions from road freight in the U.S. have increased by over 70% since 1990. While road transport remains the most significant contributor to global warming, the impact of ship and aviation transport is also frightening (Behera et al., 2019; Gupta et al., 2019; Singh & Das, 2018). According to a report released by the World Economic Forum, the future of logistics is in doubt due to the industry's detrimental impact on the environment (Das & Mondal, 2016; Jain et al., 2018; Mondal et al., 2017).

Following the Paris Climate Change Agreement, the report states that the industry needs to develop sustainable and responsible approaches to using new technologies such as drones and self-driving cars. Unfortunately, the report only addresses one of the industry's most critical issues. According to a study conducted by Accenture, digital transformation could help reduce the logistics industry's greenhouse gas emissions by up to 3.6 billion metric tons by 2025. However, they noted that the link between economic growth and greenhouse gas emissions remains unclear (Das et al., 2022a). Without proper business practices, a global gap of 8 billion tonnes will develop between the demand and supply of natural resources by 2030. This issue could cost the global economy around $4.5 trillion by 2030. It is not only about ethics but also about the economic impact of sustainability. As more and more companies realize the importance of this issue, they are starting to make it a priority (Das et al., 2022b). Unfortunately, many of them fail to follow through with their strategic actions. Despite the increasing number of sustainability initiatives, many are still tied to new digital technologies. One of these is the emergence of new business models that utilize circular economy principles (Mondal et al., 2022).

New platforms such as Uber Freight and Convoy are helping improve the efficiency of the supply chain by reducing the number of intermediaries. They also help reduce the industry's environmental impact by improving driver behavior and route optimization. New digital technologies such as the Internet of Things and advanced robotics are also assisting companies to enhance their collaboration with their partners across the supply chain (Ngoc-Vinh et al., 2022). It is a crucial step in reducing greenhouse gas emissions and improving the efficiency of the transport sector. One of the most effective ways to reduce the cost and improve the efficiency of the transport sector is by sharing warehouse space (Tien-Dung et al., 2022a). It allows companies to lower their total cost of ownership and reduce their carbon footprint. The Internet of Things is also helping companies improve their visibility in their supply chains by allowing them to track and trace their assets. It is also contributing to the development of new CSR initiatives (Minh-Nhat et al., 2022a).

Big data analytics and artificial intelligence can also help companies lower their freight costs. Combining machine learning and Big Data, these technologies can also help companies make better decisions and minimize risk (Nguyen et al., 2022). For instance, in the maritime logistics industry, data to improve vessel speeds and reduce waiting times can help decrease the time it takes to reach ports.

Green Initiatives of the European Commission

The European Commission's Green Deal aims to address the various environmental issues related to digital transformation. These issues are as follows.

a. The digital transformation can help create a circular economy by contributing to reducing greenhouse gas emissions and improving the collection and recycling of electronic waste. It is done through technological advancement that can help better recycle and reuse the materials used. The technological advances that have been made in the past few years have led to the creation of more incentive programs for consumers to recycle their electronic waste.
b. Digital technologies can help preserve and restore biodiversity. They can also help alleviate the pressures on the environment. However, their impact on the state of these services is not yet clear. For instance, collecting data on biodiversity and ecosystem services could help inform policy goals. Technology can also help visualize and communicate biological data, increase public awareness, and support policies (Minh-Nhat et al., 2022b). However, these are not enough preconditions for effective action. Instead, business models that are biodiversity-friendly should be developed to support the protection and restoration of biodiversity.
c. Smart farming is a process that aims to reduce greenhouse gas emissions and water consumption. Some quantitative assessments showed that these changes could be achieved through trial and pilot chapters. However, these findings are not yet ready to be implemented in other locations (Tien-Dung et al., 2022b). There are various potential environmental and sustainability benefits associated with enhanced traceability in the food supply chains. However, most studies on this topic focus on the economic benefits of this process.

d. Artificial intelligence and blockchain are some of the technologies that can help reduce greenhouse gas emissions and achieve a zero-pollution environment. AI-based tools have been developed to monitor and forecast pollution levels in different areas, such as traffic lights and autonomous vehicles. On the other hand, blockchain technology can be used to reward those who help reduce greenhouse gas emissions.
e. The increasing focus on citizen science has resulted in creating new opportunities for people to contribute to the protection of the environment. Digitization has the potential to improve the quality of environmental knowledge and provide more sustainable policies. It can help create and distribute relevant data at high speed, and it can also be used for new research.

A Data-Driven Approach to Sustainability

a. Due to the potential of data to inform scientific and technological innovation, political actors and scientists are calling for the establishment of open data policies. These policies aim to provide the public with access to scientific data and other relevant information. Various publications have called for the establishment of data-sharing platforms, which would allow the public to access and use data for environmental policies. Also, the increasing awareness about the value of private data has prompted more people to consider its use in scientific and technological innovation. Data collected by private companies could be used for various purposes, such as planning and implementing environmental policies. However, the exact obligations of these companies to share their data are still unclear.
b. New technologies are being used to improve the efficiency and effectiveness of environmental monitoring. One of these is blockchain and remote sensing to collect and analyze data. It could allow for faster and more accurate reporting of ecological data.
c. Through the availability of scientific data and open government data, policymakers can make informed decisions regarding the environment. It could also improve the transparency of public decision-making. Environmental organizations can also carry out their monitoring and control procedures through digital technologies.

d. Consumers can also benefit from better information about the environmental costs of their products and services by making informed decisions. In addition, digital applications such as virtual reality and gaming can help raise awareness about the environment. A data-based nudge is a potential tool for improving the efficiency and effectiveness of government. It can also be used to incentivize behavior. However, it can also raise ethical and legal issues.

ICT-Enabled Solutions Can Help Address Environmental Problems

Despite the positive effects of digitalization on the environment, it is unclear if the indirect impacts can outweigh the adverse effects of greenhouse gas (GHG) emissions. For instance, various environmental categories, such as water, biodiversity, and resource depletion, are not well studied. Supporters of digitalization often cite the potential to solve these problems. Despite the possibility of digitalization to reduce greenhouse gas emissions, the evidence of its effect on the environment is not yet clear. It is due to the various factors that affect the environment, such as consumer behavior and technological systems, which are as follows.

a. The extraction and processing of raw materials such as copper, cobalt, and gold and the production of electronic components such as integrated circuits are the main contributors to the depletion of resources globally.
b. The indirect and direct impacts of information and communications technology (ICT) on land use and biodiversity are complex and often poorly studied. Therefore, it is essential to consider the various factors that affect these two areas. The extraction of natural resources to produce hardware is one of the significant impacts of this technology. Other factors such as the improper disposal of waste from the production of electronic equipment also contribute to the depletion of resources. Other factors such as greenhouse gas emissions from power generation can also negatively impact biodiversity. Aside from greenhouse gas emissions, power generation also has environmental impacts. For instance, the development of underwater data cables has not been thoroughly studied.

c. The effects of digitalization on the environment are complex and require a comprehensive understanding of how it affects various aspects of society. Therefore, it is essential to assess the digitalization process's complexity and associated data flow increase when considering the indirect and systemic impacts. Analyzing the effects of digitalization on the physical existence of information technology (ICT) goods is a fundamental step in assessing their impact. There are still gaps in understanding the various direct implications of digitalization. For instance, there are not enough studies on the environmental impacts of data centers and data transmission networks. Also, studies on the systemic effects of digitalization have shown that they can have a wide-reaching effect. There is an urgent need for more studies on the impact of digitalization on the environment beyond greenhouse gas emissions.
d. Although digitalization can lead to various environmental benefits, it cannot be assumed to lead to a significant increase in resource consumption or energy production. A comprehensive approach is needed to understand the various impacts of digitalization and how it can be implemented sustainably. It can involve looking at the use phase and the different phases of the supply chain, and the end of life of the equipment. The implementation of digitalization must be regulated to ensure that its adverse environmental effects are minimized. It can be done through energy and resource efficiency measures.

The potential regulation of digitalization to reduce the negative environmental impacts is not widely acknowledged. There is a lack of empirical evidence on the various aspects of the effect. The literature on the multiple types of effects is also incoherent and lacks sufficient evidence on the relief potential of policies. Despite the lack of evidence, there is still enough evidence that the regulation of digitalization could reduce the non-energy impacts of the technology. Unfortunately, it is not feasible to quantify the potential reduction of environmental impacts of digitalization due to the lack of sufficient evidence. Therefore, it is important to develop policies and procedures that can help minimize the negative environmental impacts of digitalization.

The various environmental parameters needed to govern the digital transformation are grouped together into entry points designed to minimize its environmental impact. These entry points can be categorized into ICT for Green and Greening ICT. Improving product information is an entry point for the green initiative's Information and Communication Technology (ICT). Manufacturers should also provide details about their use of critical raw materials such as those used in the production of products. It can help remanufacturers and policymakers make informed decisions regarding using these materials. Sustainable ICT consumption aims to increase environmental information and social engagement. It can be achieved by guiding consumers to consider the environmental impacts of their actions and choosing sustainable solutions. One of the ways to strengthen ICT consumption is through regulations on the advertising practices of internet providers. It will help consumers make informed decisions and reduce their environmental impact. Digital technologies can help improve the management of the environment by allowing people to collect and share environmental data. These tools can also help prevent harmful activities to the environment, and they can help implement more effective environmental policies. However, they should also be supported by research to identify the effects of their use. Although these tools can be used more effectively, they should also be supported by studies to determine their use's possible environmental and social effects.

An overall framework is needed to increase the efficiency of the ICT sector and reduce its absolute resource consumption. Various options are available to achieve this goal, including adopting targets and economic incentives. Due to the immense environmental and social impacts of mining, the sector's regulation must be strengthened. It can be done by establishing effective monitoring and enforcement mechanisms in non-EU mining countries. At the same time, it is also vital that the mining industry adopt responsible behavior and promote transparency in its operations. It can be done by introducing due diligence obligations and the use of blockchain technology. Developing standards that prioritize the sustainability impacts of various supply chains is also necessary to ensure that blockchain technology can be used effectively.

The link between digitalization and the circular economy is critical. On the other hand, using energy and raw materials to produce information and communications technology (ICT) can have detrimental ecological impacts. Companies can make significant contributions to the circular economy using digital tools and data. It could involve improving production efficiency and the use of natural resources. An entry point for the producer is the original producer's responsibility for the environmental impacts of a product. This policy aims to encourage producers to make changes to reduce their waste management costs. The producer should involve making changes to the product's recyclability and reusability. It should also include engaging in other designs for environmental activities. Digital solutions can also be used to improve the efficiency of the supply chain.

The development of regulations and other policies related to the circular economy should be coordinated to ensure that digitalization and the transition to a more sustainable economy are appropriately managed. The extended lifespan of mobile information and communications technology (ICT) devices such as smartphones and tablets could lead to the recovery and remanufacturing of used materials. It could also help reduce the number of discarded devices in households. Many products that are still functional are replaced even though they are no longer helpful. The factors that influence this phenomenon include the short innovation cycles and the pricing models of service providers. One of the most effective ways to reduce the barriers to implementing the circular economy is by establishing a framework for the repair of ICT goods. It could help reduce the concerns of consumers about data privacy.

It is also essential that the collection rate of ICT goods is increased once they reach their end of life. It can be done by developing a well-functioning secondary materials supply chain. The exchange of information and the quality standards established in the secondary materials market are some of the factors needed to improve this process.

The remanufacturing of critical raw materials (CRM) is a complex process that involves the use of embedded CRMs. Manufacturers must become more aware of these components and their locations to plan for recovery. Efforts to reduce the amount of e-waste are needed to build a more circular economy. It can be achieved through the implementation of policies and the use of recycling technology. The European Union should also encourage the collecting and recycling of e-waste in developing countries. It could be done by establishing an international fund that would provide pre-defined premium payments for soundly recycled e-waste. The use of ultrapure chemicals in the ICT industry is a significant concern for the environment. To better understand the environmental impacts of these chemicals, the transparency of these chemicals should be increased. Greening data center operation: Various initiatives such as using waste heat can be carried out to reduce the data center's energy consumption.

One of the first steps that can be taken to promote sustainable software is by establishing a voluntary program that would allow software developers to apply for a label that would indicate their commitment to reducing their software's resource consumption. Complex algorithms are used in various digital applications, such as search engines and trading platforms to determine the best routes for autonomous driving. Although the criteria used to determine these decisions are transparent, they still fulfill their duties. These algorithms' complexity and decisions can have wide-ranging negative impacts on the environment. Due to the increasing importance of data, the debate about the proper use of data is becoming more relevant to environmental policies. This paper does not provide a comprehensive list of regulatory alternatives. Instead, it focuses on the general importance of data and the various arguments that can be made regarding its rights. The paper also addresses the different opinions that can be made regarding the right to access data.

There are limitations in the current literature on the effects of digitalization on the environment. Data gaps and blind spots are also identified in the evidence.

a. We are developing methods to measure and assess the impacts of climate change on biodiversity. These include evaluating the effects of complex cause-effect chains.
b. This chapter aims to promote the use of life cycle assessments beyond the traditional energy and greenhouse gas aspects of network infrastructure.
c. This chapter aims to promote the use of life cycle assessments for new technologies and their environmental impacts.

The authors mainly intend to discuss certain aspects of sustainable development goals related to environment protection and supply chain management.

a. This chapter aims to develop a set of life cycle assessments that are easily accessible and geared toward assessing new technologies' business models and digital applications. The chapter seeks to expand the scope of impact categories in environmental assessments. These include the effects of new technologies on biodiversity and the environment. Case studies are being conducted to analyze the impacts of information and communications technology (ICT)-enabled solutions on the environment. These studies aim to identify sectors and applications that are expected to have more environmental benefits than their risks. These studies should review the state of the current research on these topics and develop new methodologies.

b. The chapter aims to address the data gap related to the chemicals used in the ICT and semiconductor industries. It will also establish a standard set of secondary databases that will allow users to access and compare the results of environmental assessments easily. The chapter will also develop standards that will enable Life Cycle Inventories to collect and manage data related to energy and material flows. It will help them improve their databases and provide a more accurate and reliable view of their operations.
c. The chapter aims to identify the significant points of sustainable ICT consumption and how consumers can be educated about the environmental impact of their actions. It includes the activities that they perform through the use of ICT hardware and applications. It is also essential to study the attitudes and knowledge of consumers regarding the various sustainable consumption options available in the ICT goods and applications market. It will help inform the development of effective programs and policies that can help consumers make informed decisions.
d. The chapter also aims to identify policies and institutions that can contribute to the sustainable development of digital technology. It includes regulations related to the use of Big Data. Despite the increasing prominence of the environmental impacts of new technologies, these have not been considered in legal policy debates.

The European Green Deal is linked to digital transformation. It calls for establishing a comprehensive digital strategy and the development of digital instruments. The Commission will adopt an industrial approach in 2020 that will address the challenges of digital change and environmental sustainability. This strategy is expected to play a critical role in achieving the goals of the Green Deal.

The Green Deal acknowledges the importance of digitalization in developing the energy system. It is also expected that the integration of the European market will contribute to the decarbonization of the transportation system. The Commission will also investigate measures to accelerate digital technologies in addressing climate change and protecting the environment. The Green Deal acknowledges the importance of data-driven innovation. It will allow organizations to make informed decisions and improve the efficiency of their operations by leveraging the power of data. New opportunities will be created using digital platforms and artificial intelligence to monitor and control the environment. The Green Deal also acknowledges the importance of digital technologies in addressing the environment. It calls for establishing a comprehensive digital strategy and the development of digital instruments. The Commission will also investigate measures to improve the energy efficiency of the ICT sector.

Approaches Adopted for This Study

The study's goal is to review the literature and identify the most relevant studies that deal with the various aspects of sustainability. These include environmental assessments, life cycle assessments, and carbon footprint studies. We have limited the study's scope to studies published after 2010. The information collected is based on a conceptual framework that defines information and communications technology (ICTs) in the natural environment. It divides the impact into three categories: direct, indirect, and systemic.

The direct impacts of information and communications technology are those caused by the use and disposal of products. The second-order effects of ICTs reduce the environmental impact of various activities. They can make production and consumption more resource-efficient. However, these effects can also have adverse effects on the environment. The third-order impact of ICTs is those triggered by the transition to a digital society. They involve the impact of various non-technological factors, such as the changes in consumer behavior. Systemic consequences are those that are expected to have long-term effects. Despite the widely accepted classification of impact, it is not always easy to apply this concept to the various scientific questions related to its nature. For instance, the variety of desirable and adverse effects is not yet established. Another limitation of this concept is that it only considers the complex cause-effect chains of ICTs. For instance, studies on the rebound effects have only just started to be conducted.

The use of information and communications technology (ICT) can significantly impact the efficiency of various processes, such as industrial production and transport. This environmental benefit is also reflected in the frequency and duration of these processes' use. The digitization of processes has broad-ranging effects, transforming the way we do business, and providing us with more information about our environment. With vast amounts of data, it can give us deeper insight into our society. Using technology can also transform the way we interact with the world around us. It can provide us with powerful tools and techniques to help us make informed decisions and improve our environment. Despite the various advantages of digitalization, there is currently not enough evidence supporting the concept of ecological relief from its effects. Therefore, it is essential to keep in mind that there are still many uncertainties about its implementation.

The use of information and communication technology (ICT) has various environmental benefits. It can help improve the efficiency of multiple processes, such as industrial production and transport. The frequency and duration of the use of these processes are also considered. The use of technology has various systemic effects, such as transforming the way we consume services and improving our processes' efficiency. It can also boost our knowledge about the environment and our behavior. Using technology, we can also make the world more sustainable by automatically and effectively influencing its physical and social conditions. New tools and techniques can help improve the efficiency of various processes and create new business models. Unfortunately, there is currently not enough evidence supporting the concept of ecological relief from digitalization. The effects of digital solutions are known to have various uncertainties and potential adverse effects.

Supply Chain Trends for Post-Pandemic Recovery and Sustainability

Today, supply chain management is more than just a function that provides cost-control solutions. It's also a strategy that can help drive revenue growth. By analyzing the entire supply chain global trends, supply chain managers can improve their efficiency and mitigate risk (Attaran, 2020). New directions for post-2020 are helping drive the evolution of the end-to-end supply chain, a comprehensive view of the various activities involved in delivering goods and services to the consumer. According to the Global Supply Chain Institute, the end-to-end supply chain is an integrated system that enables companies to provide products and services from their supplier to their consumers (Cagliano et al., 2021). The supply chain is the path that connects various resources to users. Changing geopolitical forces, climate, and technology can affect this process. Over the last five years, the evolution of supply chain management systems has allowed organizations to transform their operations into opportunities (Chouki et al., 2020). Some of these opportunities are as follows;

a. The year 2020 has been full of innovation. According to Gartner, eight new technology trends emerged in 2019. One of these is blockchain, which could help address the challenges of fraud, visibility, and efficiency play.
b. The next generation of supply chain technology is called wise. It combines artificial intelligence, big data, and the Internet of Things. Predictive analytics can also help organizations improve their operations ("Control Supply Chain Risks in Digital Transformation," 2022). Digitalization is a trend that can transform the way value is created through technology. For instance, by embedding sensors in the supply chain, a restaurant can now monitor the movement of its fish.
c. The global supply chain is constantly evolving due to various factors. For instance, materials are often sourced from different countries and transported to a third country (Creazza et al., 2021). It means that the supply chain has to react quickly to address the changes brought about by global trade.
d. Due to the constant changes in the global business environment, risk management has become a vital part of every organization's operation (Guo et al., 2022). It is responsible for managing the risks associated with the supply chain to minimize delays and minimize disruptions.
e. Due to the increasing importance of social and environmental issues, business strategy has become more focused on sustainability. This concept involves considering the various factors that affect the environment and developing sustainable methods and practices (Gupta et al., 2020). Doing so can help companies lower their costs and ensure that their supply chains remain operational.

The key to success in these areas is supply chain visibility. Without it, innovation will not thrive, and collaboration will be lost. Therefore, organizations must have the necessary tools and resources to enable their visibility. Having visibility into the various parts of the supply chain network is also essential to ensure that a company can manage risk and be sustainable (Muafi & Sulistio, 2022). It is done through continuous monitoring and analysis of the various aspects of the supply chain. Supply chain management trends are very much in practice for post-pandemic recovery. Despite the various trends that are being talked about in 2020, sustainability and technology are still the two most talked about trends in 2020 (Preindl et al., 2020). These two factors can produce positive outcomes in the transportation and logistics sectors.

a. Rethinking data: With the rise of machine learning, companies can now manage their data at the data level. New technology is now more flexible, which means it can be used to meet the varying needs of different companies. It eliminates data segregation and allows them to connect their various systems. It will allow them to share their data safely and securely.
b. In technology stacking, managers can create new capabilities using multiple technologies. It allows them to customize their software to meet their specific needs. This approach is like how warehouse management systems combine various systems to deliver personalized results. The cost of implementing technology stacking depends on a company's size and how it customizes its solutions. According to researchers, there are two types of stacking: the first focuses on continuous improvement, and the second uses new capabilities ("Supply Chain Management Concept and Digital Economy: Digital Supply Chain Technological Innovation," 2021). With the availability of various stacking options, companies can now customize their software to meet their specific needs. It will allow them to compete more effectively (Wieland, 2020).
c. The rise of digitalization is transforming the way companies do business, with new technologies that can help them improve their supply chains and create new forms of value. It is done by using new tools and techniques that can help improve the efficiency of their operations. New supply chain management software can help improve the efficiency of your operations and generate new revenue streams.
d. The rise of digitalization is transforming the way companies do business, with new technologies that can help them improve their supply chains and create new forms of value. One of these is blockchain, which can help companies manage their contracts. Despite the availability of collaborative software, many supply chain managers still rely on email and excel in managing their short-term projects.

Digitalization can help improve the efficiency of the entire supply chain by allowing them to connect and collaborate across different departments.

e. The rise of new capabilities and processes within the supply chain management software is driving the need for more innovation. However, creativity and technical expertise are the only factors keeping the new models from being fully realized.
f. Despite the increasing complexity of today's supply chains, they still need to maintain their human capital to keep up with the changes brought about by new technologies and automation. To be successful, supply chain managers need to be digitally savvy and have the right skills to match their companies' needs.
g. The nature of work has changed over the years, and it's only going to get more complex in the next two decades. According to a survey conducted by ManpowerGroup, the rise of digitalization and automation will lead to a net gain in employment.

Sustainable Trends for Digitalized Supply Chain Management

The rise of digitalization of the supply chain is creating new opportunities for companies to improve their bottom line through internal collaboration. This process can be achieved with new technologies that enable supply chain managers to develop new forms of value. New supply chain management software can help drive revenue-driving results for companies. Supply chain managers can now start implementing these new technologies. As technology continues to transform the way we operate, so does how we think about sustainability in the supply chain. The Global Supply Chain Institute discusses the latest supply chain innovations driving positive change. Due to the complexity of the supply chain, managers must adopt sustainable practices. According to GSCI, these can provide long-term financial returns and reduce environmental impacts. The most sustainable companies talked about the various steps to improve their operations and reduce their environmental impact. These steps are as follows;

a. With sustainability becoming a significant business driver, many companies are starting to make it a part of their operations.

According to data, many millennials are looking for companies committed to making a positive impact on the environment.

b. Top companies have a clear vision and strategy for sustainability. It can be done by developing a culture of transparency, where employees are expected to understand how their actions can contribute to a more sustainable environment.
c. Getting involved with sustainable suppliers is a vital step for any company that wants to become a part of the solution to the environment.
d. One of the most critical steps in establishing a culture of transparency is to ensure that all of your products are traceable. It can help prevent harmful practices and ensure that the materials used are produced according to the correct environmental standards.
e. Using data to improve your company's performance is another step that can be taken to create a sustainable supply chain. By continuously monitoring and analyzing their operations, companies can improve their environmental impact.
f. Getting involved with experts: Working with experts is also vital in establishing a successful supply chain. One of the most important steps that a company can take is to partner with a supply chain institute. Through collaboration, companies can develop effective strategies and solutions.

Conclusion

According to various studies, companies that can improve their efficiency and supply the goods and services demanded by their customers will be the leaders of their industries. The time has come for companies to adopt a new sustainability mindset in their supply chain operations. This new mandate will require them to embed sustainability into their digital technologies. We have been using technology to collect and analyze sustainability data for years. However, due to the limited number of tools available, we have had to rely on agents and traders to share this data. This new skill set will allow companies to transform their supply chains and improve efficiency. They will need to implement various digital tools to visualize and analyze their sustainability data, create interoperable systems, and use augmented reality to enhance worker engagement. By implementing digital tools and techniques, companies can improve their efficiency and create new opportunities for innovation.

References

Attaran, M. (2020). Digital technology enablers and their implications for supply chain management. *Supply Chain Forum: An International Journal, 21*(3), 158–172. https://doi.org/10.1080/16258312.2020.1751568

Behera, M., Dash, M., Mohanty, A. K., & Das, S. (2019). Entrepreneurial ecosystem: A geographical upliftment. *Revista Espacios, 40*, 26–36. http://www.revistaespacios.com/a19v40n01/19400126.html

Cagliano, A. C., Mangano, G., & Rafele, C. (2021). Determinants of digital technology adoption in supply chain. An exploratory analysis. *Supply Chain Forum: An International Journal, 22*(2), 100–114. https://doi.org/10.1080/16258312.2021.1875789

Chouki, M., Fernandes, V., Ruel, S., & Borja De Mozota, B. (2020). Towards the «Digital Project Integration»: Contributions of an analogical approach with the Supply Chain Integration. *Supply Chain Forum: An International Journal, 21*(3), 219–228. https://doi.org/10.1080/16258312.2020.1752608

Control Supply Chain Risks in Digital Transformation. (2022). *Journal of Organizational and End User Computing, 34*(7). https://doi.org/10.4018/joeuc.298680

Creazza, A., Colicchia, C., Spiezia, S., & Dallari, F. (2021). Who cares? Supply chain managers' perceptions regarding cyber supply chain risk management in the digital transformation era. *Supply Chain Management: An International Journal, 27*(1), 30–53. https://doi.org/10.1108/scm-02-2020-0073

Das, S. (2020a). *Media impact of advertising on consumer buying behaviour a comparative study of different media*. Shiksha o Anusandhan University. http://hdl.handle.net/10603/273691

Das, S. (2020b). Innovations in digital banking service brand equity and millennial consumerism. In *Digital transformation and innovative services for business and learning* (pp. 62–79). IGI Global. https://doi.org/10.4018/978-1-7998-5175-2.ch004

Das, S. (2021a). *Digital entertainment: The next evolution in service sector*. Springer Nature.https://doi.org/10.1007/978-981-15-9724-4. Print ISBN 978-981-15-9723-7. Online ISBN 978-981-15-9724-4

Das, S. (2021b). *Search engine optimization and marketing: A recipe for success in digital marketing* (1st ed.). CRC Press. https://doi.org/10.1201/9780429298509

Das, S. (2021c). A systematic study of integrated marketing communication and content management system for millennial consumers. In *Innovations in digital branding and content marketing* (pp. 91–112). IGI Global. https://doi.org/10.4018/978-1-7998-4420-4.ch005

Das, S., & Mondal, S. R. (2016). *Yours digitally: Patanjali*. Available at SSRN 2737299.

Das, S., Mondal, S. R., & Sandhu, K. (2022a). Music logos drive digital brands: An empirical analysis of consumers' perspective. *Journal of Strategic Marketing*, 1–16. https://doi.org/10.1080/0965254X.2022.2098526

Das, S., Mondal, S., Puri, V., & Vrana, V. (2022b). Structural review of relics tourism by text mining and machine learning. *Marketing, 8*(2), 25–34. https://doi.org/10.5281/zenodo.7358349

Duman, Ü. O., & Das, S. (2021). The dynamic rise of digital brands' market mavens in digital entertainment: A complete know how for curious customers. In *Digital entertainment* (pp. 147–172). Palgrave Macmillan. https://doi.org/10.1007/978-981-15-9724-4_8

Duy, N. T., Mondal, S. R., Van, N. T. T., Dzung, P. T., Minh, D. X. H., & Das, S. (2020). A study on the role of web 4.0 and 5.0 in the sustainable tourism ecosystem of Ho Chi Minh City, Vietnam. *Sustainability*, *12*(17), 7140. https://doi.org/10.3390/su12177140

Guo, X., Xia, W., Feng, T., & Sheng, H. (2022). Sustainable supply chain finance adoption and firm performance: Is green supply chain integration a missing link? *Sustainable Development*. https://doi.org/10.1002/sd.2307

Gupta, D. K., Jena, D., Samantaray, A. K., & Das, S. (2019). HRD climate in selected public sector banks in India. *Revista ESPACIOS*, *40*(11).

Gupta, S., Modgil, S., Gunasekaran, A., & Bag, S. (2020). Dynamic capabilities and institutional theories for Industry 4.0 and digital supply chain. *Supply Chain Forum: An International Journal*, *21*(3), 139–157. https://doi.org/10.1080/16258312.2020.1757369

Jain, S., Jain V., & Das, S. (2018). Relationship analysis between emotional intelligence and service quality with special evidences from Indian banking sector. *Revista ESPACIOS*, *39*(33).

Minh-Nhat, H. O., Majerova, J., & Das, S. (2022a). Taking steps to help the phoenix rise from ashes: A roadmap for digital enterprises to develop new sustainable goals. In *Sustainable development and innovation of digital enterprises for living with COVID-19* (pp. 35–51). Springer. https://doi.org/10.1007/978-981-19-2173-5_3

Minh-Nhat, H. O., Nguyen, H. L., & Mondal, S. R. (2022b). Digital transformation for new sustainable goals with human element for digital service enterprises: An exploration of factors. In *Sustainable development and innovation of digital enterprises for living with COVID-19* (pp. 85–103). Springer. https://doi.org/10.1007/978-981-19-2173-5_6

Mohanty, P. C., Dash, M., Dash, M., & Das, S. (2019). A study on factors influencing training effectiveness. *Revista Espacios, 40*, 7–15. http://www.revistaespacios.com/a19v40n02/19400207.html

Mondal, S. R. (2020a). *Factors influencing store image loyalty and satisfaction from customer perspective an empirical study in retail sector*. Shiksha o Anusandhan University. http://hdl.handle.net/10603/273688

Mondal, S. R. (2020b). A systematic study for digital innovation in management education: An integrated approach towards problem-based learning in Vietnam. In *Digital innovations for customer engagement, management, and organizational improvement* (pp. 104–120). IGI Global. https://doi.org/10.4018/978-1-7998-5171-4.ch006

Mondal, S. R. (2021). A systematic study of new age consumer engagement and exploration for digital entertainment for over-the-top platforms in various digital media. In *Innovations in digital branding and content marketing* (pp. 113–133). IGI Global. https://doi.org/10.4018/978-1-7998-4420-4.ch006

Mondal, S. R., Das, S., Musunuru, K., & Dash, M. (2017). Study on the factors affecting customer purchase activity in retail stores by confirmatory factor analysis. *Revista ESPACIOS, 38*(61).

Mondal, S. R., & Das, S. (2021). Examining diabetic subjects on their correlation with TTH and CAD: A statistical approach on exploratory results. In *Machine Learning and the Internet of Medical Things in Healthcare* (pp. 153–177). Academic Press. https://doi.org/10.1016/B978-0-12-821229-5.00007-0

Mondal, S. R., Majerova, J., & Das, S. (2022). *Sustainable development and innovation of digital enterprises for living with COVID-19*. Springer. https://doi.org/10.1007/978-981-19-2173-5

Mondal, S. R., Virgilio, F. D., & Das, S. (2022). *H.R. analytics and digital H.R. practices*. Springer. https://doi.org/10.1007/978-981-16-7099-2

Muafi, M., & Sulistio, J. (2022). A nexus between green intelectual capital, supply chain integration, digital supply chain, supply chain agility, and business performance. *Journal of Industrial Engineering and Management, 15*(2), 275. https://doi.org/10.3926/jiem.3831

Nadanyiova, M., & Das, S. (2020). Millennials as a target segment of socially responsible communication within the business strategy. *Littera Scripta, 13*(1), 119–134. https://doi.org/10.36708/Littera_Scripta2020/1/8

Ngoc-Vinh, N., Tien-Dung, P., & Das, S. (2022). Digitization of business; need for recombination during COVID-19. In *Sustainable development and innovation of digital enterprises for living with COVID-19* (pp. 1–17). Springer. https://doi.org/10.1007/978-981-19-2173-5_1

Nguyen, T. D. M., Mondal, S. R., & Das, S. (2022). Digital entrepreneurial transformation (DET) powered by new normal sustainable developmental goals (n-SDGs): Elixir for growth of country's economy. In *Sustainable development and innovation of digital enterprises for living with COVID-19* (pp. 69–84). Springer. https://doi.org/10.1007/978-981-19-2173-5_5

Preindl, R., Nikolopoulos, K., & Litsiou, K. (2020). Transformation strategies for the supply chain: The impact of industry 4.0 and digital transformation. *Supply Chain Forum: An International Journal, 21*(1), 26–34. https://doi.org/10.1080/16258312.2020.1716633

Ravi, S., & Mondal, S. R. (2021). Digital entertainment based do it yourself content & advertisement as a factor of driving force for trust & customer patronage. In *Digital entertainment* (pp. 127–146). Palgrave Macmillan. https://doi.org/10.1007/978-981-15-9724-4_7

Sharma, E., & Das, S. (2020). Measuring impact of Indian ports on environment and effectiveness of remedial measures towards environmental pollution. *International Journal of Environment and Waste Management, 25*(3), 356–380. https://doi.org/10.1504/IJEWM.2019.10021787

Sharma, E., & Das, S. (2021). Integrated model for women empowerment in rural India. *Journal of International Development*, 1–18. https://doi.org/10.1002/jid.3539

Sharma, E., Nigam, N., & Das, S. (2020). Measuring gap in expected and perceived quality of ICT enabled customer services: A systematic study of top ten retailers of India. *International Journal of Applied Systemic Studies, 9*(2), 159–184. https://doi.org/10.1504/ijass.2020.113260

Singh, L. B., Mondal, S. R., & Das, S. (2020). Human resource practices & their observed significance for Indian SMEs. *Revista ESPACIOS*, *41*(07). http://www.revistaespacios.com/a20v41n07/20410715.html

Singh, S., & Das, S. (2018). Impact of post-merger and acquisition activities on the financial performance of banks: A study of Indian private sector and public sector banks. *Revista Espacios Magazine, 39*(26), 25.

Siri, R., & Das, S. (2021). A study on processing of information storage & use of new age consumers in digital wellness sector through story telling & creating interest. In *Digital entertainment* (pp. 45–63). Palgrave Macmillan. https://doi.org/10.1007/978-981-15-9724-4_3

Siri, R., Mondal, S. R., Das, S. (2020). Hydropower: A renewable energy resource for sustainability in terms of climate change and environmental protection. In *The handbook of environmental chemistry*. Springer. https://doi.org/10.1007/698_2020_635

Supply Chain Management Concept and Digital Economy: Digital Supply Chain Technological Innovation. (2021). *Indian Journal of Economics and Development*, 928–933. https://doi.org/10.35716/ijed/21272

Tien-Dung, P. H. A. M., Majerova, J., & Das, S. (2022a). Phases of possible recovery of digital enterprises in new normal business for living with COVID-19 times: Opportunities for a new era in sustainable development goals. In *Sustainable development and innovation of digital enterprises for living with COVID-19* (pp. 19–33). Springer. https://doi.org/10.1007/978-981-19-2173-5_2

Tien-Dung, P., Nguyen, X. D., & Das, S. (2022b). Sustainable goal achievement by digital revolution during and after pandemic; how much one wins and losses: A bird's eye view for future planning. In *Sustainable development and innovation of digital enterprises for living with COVID-19* (pp. 133–147). Springer. https://doi.org/10.1007/978-981-19-2173-5_8

Tiwari, M., & Mondal, S. R. (2022). Technological dynamism of workforce management for effective education service delivery during and after Covid-19. In *H.R. analytics and digital H.R. practices* (pp. 199–215). Palgrave Macmillan. https://doi.org/10.1007/978-981-16-7099-2_8

Van, N. T. T., Vrana, V., Duy, N. T., Minh, D. X. H., Dzung, P. T., Mondal, S. R., & Das, S. (2020). The role of human-machine interactive devices for post-COVID-19 innovative sustainable tourism in Ho Chi Minh City, Vietnam. *Sustainability, 12*(22), 9523. https://doi.org/10.3390/su12229523

Wieland, A. (2020). Dancing the supply chain: Toward transformative supply chain management. *Journal of Supply Chain Management, 57*(1), 58–73. https://doi.org/10.1111/jscm.12248

Yegen, C., & Mondal, S. R. (2021). Sharenting: A new paradigm of digital entertainment of new age parenting and social media. In *Digital entertainment* (pp. 213–231). Palgrave Macmillan. https://doi.org/10.1007/978-981-15-9724-4_11

CHAPTER 9

Digitalization of Enterprises from a New Normal Sustainable Development Perspective: Opportunities and Challenges

Francesca Di Virgilio and Subhankar Das

Introduction

The rise of digital connectivity has transformed the way people communicate and conduct business. It has become a vital component of maintaining social and economic activities in the face of global isolation (Mondal et al., 2022). The lack of preparation for the surge of demand has caused various factors such as network congestion and the decline in Internet speed to affect the quality of service in some countries (Tiwari & Mondal, 2022). Unequal access to broadband connectivity can lead to

F. Di Virgilio
Department of Economics, University of Molise, Campobasso, Italy
e-mail: fradivi@unimol.it

S. Das (✉)
The Honors Programme, Southern Star Institute of Management, Duy Tan University, Da Nang, Vietnam
e-mail: subhankardas@duytan.edu.vn

S. R. Mondal et al. (eds.), *New Normal in Digital Enterprises*,
https://doi.org/10.1007/978-981-19-8618-5_9

social inequality and undermine stability. During the COVID-19 period, many countries experienced network congestion (Mondal & Das, 2021). Various factors affected the quality of service. The use of the network during daytime hours in residential areas caused congestion in the last mile networks (Sharma & Das, 2021). This issue can affect the availability of various high-bandwidth services such as video and cloud services.

The key to successful innovation is collaboration. Over the past couple of years, the manufacturing industry has shown that it can still face new challenges and opportunities by working together (Das, 2021a). Despite the situation's complexity, there are still common factors that affect every company's operation. When the global pandemic hit, many companies had to adapt their working models to address the changes brought about by the outbreak (Siri & Das, 2021). They also had to manage their supply chains and operations. The concept of digital transformation is a common framework for recovery in the manufacturing industry (Ravi & Mondal, 2021). However, different organizations are going through different phases of their journey. For instance, after experiencing the initial crisis, they need to be resilient before starting to recover (Duman & Das, 2021). Working together to implement digital transformation is a common framework that manufacturers can use to accelerate their recovery. For instance, during a recession, digital transformation aims to reduce operational costs and improve sales. Although digital solutions can help improve the efficiency of operations, they should also be focused on areas where they can make the most immediate and cost-effective changes (Yegen & Mondal, 2021).

Digitization is a process that enables manufacturers to adapt to new market demands and improve their operations (Das, 2021b). It can be done through collaboration with other organizations and with the help of data transparency. The increasing number of companies pushing for digital agility has created a convergence of the various technology platforms and operations (Sharma et al., 2020). It helps the front-line staff members perform their duties more effectively (Mondal, 2020a). The goal of digital transformation is to improve people's productivity and support their well-being (Das, 2020a). It can also enhance their everyday experiences using predictive maintenance and flexible production capabilities (Das, 2021c).

As manufacturers prepare for the next normal, they need to learn how to manage their pandemics and prevent them from happening again (Mondal, 2021). This process involves working together to develop and implement effective solutions (Siri et al., 2020). Companies need to be honest and open with themselves to create a strategy that will work (Duy et al., 2020; Van et al., 2020). They also suggest using data-driven operations to gather and analyze information. One of the most significant changes in the next normal is the shift from optimizing to protection (Singh et al., 2020). This strategy involves working together to create closed-loop ecosystems that can help reduce their environmental impact (Sharma & Das, 2020). The U.N. has set a target of achieving Net Zero emissions by 2050. To reach this target, manufacturers need to develop intelligent, digital innovations that will help them meet the goals (Das, 2020b; Mondal, 2020b). Sustainability can only be achieved through a coordinated approach between the various companies in the industry. Instead of focusing on one area, each company can focus on developing and improving its capabilities (Mohanty et al., 2019; Nadanyiova & Das, 2020).

This digitalization platform brings together people from different backgrounds and industries to create a more collaborative environment for innovation. One of the most important factors that people need to consider when it comes to collaboration is the opportunity to share their ideas and concerns with others (Behera et al., 2019; Gupta et al., 2019; Singh & Das, 2018). It can help create a more collaborative environment and avoid repeating the disruption we experienced during the pandemic. Some companies are also planning on launching or expanding their direct-to-consumer channels (Das & Mondal, 2016; Jain et al., 2018; Mondal et al., 2017). Due to the rise of e-commerce, customers' expectations have increased regarding the quality of their logistics services. They want their orders to be delivered as quickly as possible. The rising cost of health is directly linked to the number of transported goods. If that weren't enough, 40% of Europeans would experience traffic noise levels of over 55 D.B. Around 21% of all transport trips in the European Union are empty. About 30% of the truckload capacities are still unused.

The European Union unveiled its 2020 sustainability goals. These include reducing greenhouse gas emissions by 20% and improving energy efficiency by 20%. The future of transportation is being built on technology innovations, such as self-driving cars and alternative fuels. These advancements will help reduce greenhouse gas emissions and improve the safety of drivers. Despite the progress being made, the future of transportation still has a long way to go before it can be fully realized (Das et al., 2022a). Autonomous trucks are still in their early stages, and alternative fuels such as natural gas and bioethanol are still in their infancy.

The world's population is expected to reach 9 billion by 2050, which means that e-commerce will continue to grow. The market for online goods continues to grow. One of the main factors that drive this is the increasing number of consumers who expect same-day delivery on most of their purchases (Das et al., 2022b). This demand is forcing more companies to adopt new business models geared toward optimizing their supply chains. It is also beneficial for logistics providers as the middle class expands. It is causing an increase in transport infrastructure and environmental pollution (Mondal et al., 2022).

The freight transport industry can play a leading role in reducing greenhouse gas emissions and improving the efficiency of its operations. Unfortunately, the industry has not adequately addressed the issue of sustainability. The road freight transport industry in Europe is highly fragmented. Only a few large companies account for about 10% of the market share. Most companies that transport goods by road are small and medium-sized enterprises. The fragmented nature of the road freight transport industry makes it harder for the industry to meet its climate goals (Ngoc-Vinh et al., 2022). It also limits the scope of its potential to reduce greenhouse gas emissions. Many small and medium-sized enterprises (SMEs) rely on outdated technology to manage their transportation logistics operations (Tien-Dung et al., 2022a). They often lack the necessary tools and resources to improve their efficiency. Aside from the telephone and fax machines, many of these companies also use paper-based processes. Due to the emergence of digital business models, the road freight transport industry is expected to undergo a significant shift (Minh-Nhat et al., 2022a). The increasing number of online transactions will create new competitive advantages for companies and consumers.

Sustainable Development and Its Recovery Mediums

Creating jobs is one of the critical actions governments can take to promote sustainable recovery. However, it is also essential to balance the need for growth with the need for employment. Many countries in the region are taking measures to boost domestic demand to minimize the impact of the COVID-19 pandemic.

The International Labor Organization (ILO) has urged governments to develop robust social policies and employment measures to support the recovery. These include investing in infrastructure and providing social protection for the most vulnerable. Several countries in the region, including Brazil, Chile, and The Bahamas, have been implementing measures to support their companies' liquidity by reducing or delaying the payment of labor taxes. Despite the recovery, it is still not enough to simply return to business. Countries must adopt structural changes that allow them to produce and consume more energy-efficient products and services to maintain a sustainable recovery.

There are also winners and losers from the transition. For instance, the elimination of about 7.5 million jobs in the agriculture and energy sectors would cause a significant decline in the number of jobs in these areas (Nguyen et al., 2022). However, creating over 20 million new jobs in other sectors would benefit the most vulnerable individuals.

Role of Digital Enterprise

The role of small and medium enterprises (SMEs) in economic recovery is significant. More than 10 million companies are more than 30% of the GDP in the region. Due to the pandemic, over 42,000 SMEs have closed in Argentina. Companies' business models must include sustainability principles and guidelines to move toward sustainable development. It can help protect and strengthen SMEs and promote economic innovation. Digitalization is also a critical issue that needs to be addressed to boost the region's economy. Due to the various confinement measures in place in most countries, companies must adopt new technology. In Brazil, for instance, the government is providing various support programs and services to help small and medium-sized businesses (SMEs) adopt digital technology. These include the establishment of e-commerce platforms and the provision of financial services.

The importance of supporting businesses is also highlighted in the eighth Sustainable Development Goal. One of the critical actions that the government can take to boost the number of green SMEs is by providing them with targeted financing (Minh-Nhat et al., 2022b). It can help them develop their sustainability strategies and become more resilient in the face of climate change and other natural disasters. For instance, the Inter-American Development Bank approved a loan of almost US$20 million to help Honduras' tourism sector recover from the effects of the crisis.

The social protection provided by the government helps stabilize the economy and prevent further job losses. During the pandemic, the payment chain was affected, which led to the loss of employment contracts. The protection of social and economic assets during the crisis is critical to reducing the risk of economic and social damage. It is also a call to action for policymakers and governments to implement a comprehensive social protection system. A complete social protection system can help countries protect their citizens from the effects of the crisis (Tien-Dung et al., 2022b). It can provide them with various benefits such as pensions, unemployment insurance, and health care. The recovery from the problem is starting to create a sustainable future for the region. Many countries are already taking the necessary steps to reduce greenhouse gas emissions and create a more sustainable future. For instance, Costa Rica is working toward a net-zero emissions economy through its National Decarbonization Plan. The region's governments should be able to implement structural changes in their economic recovery packages to increase resilience to future crises. Social dialogue is also critical to help resolve the problem and ensure that workers and communities are treated fairly.

Due to the climate crisis, the fashion industry is shifting its focus to a more sustainable approach. Although the term sustainability may not immediately come to mind, it is essential to note that it means the quality of being able to endure for a long time (Abdallah, 2017). It is what makes luxury and fashion so unique. The planetary emergency facing humanity and the Earth is complex. It involves multiple interrelated issues such as climate change, pollution, and land degradation (Agostini et al., 2019). Despite the challenges of changing habits, a transition to a circular economy is possible. The fashion industry is a significant contributor to the global economy. It is estimated that it contributes around 2 to 8% of greenhouse gas emissions and is responsible for approximately 20% of industrial wastewater. It also has a significant impact on the environment. Despite the industry's efforts to reduce its impact on the environment, it is still not yet able to achieve carbon neutrality or environmental sustainability (Astafyeva, 2021). Therefore, the fashion industry must rethink its approach and adopt a comprehensive approach to achieve these goals. Big data and analytics can help businesses make informed decisions and improve efficiency. These tools can help them monitor and analyze the markets and identify opportunities (Irawan, 2020). Developing sustainable products and processes can also help companies reduce their environmental impact. Artificial intelligence can help reduce waste and improve the efficiency of the supply chain by analyzing data and forecasting trends. Virtual reality can also help customers try on clothes and reduce their carbon footprint (Ivanov et al., 2021). Blockchain technology can help companies track their raw materials and ensure that their practices are responsible.

We can create eco-friendly and sustainable clothes with eco-design techniques, such as 3D printing. Social media campaigns can also help change the way people think about fashion (Laksmanawati & Yuniawan, 2021). During a panel discussion on sustainability in the fashion industry, Alok Sharma, the President of the U.N. Climate Change Conference, said that the industry needs to make a fundamental shift to address the issue of climate change. He also called on the entire sector to adopt the goals of the Paris Climate Agreement. On September 30, the Italian city of Milan hosted the Pre-COP, which is the final meeting before the Conference of the Parties (COP) 26 on climate change takes place in November 2021. It aims to advance the goals of the Paris Agreement by increasing international collaboration and strengthening adaptation. The U.N. Development Program (UNDP) launched a campaign to raise the sustainability goals of the fashion industry in the lead-up to the next climate conference.

On October 7, UNDP will host a live discussion on Instagram with Federico Marchetti, the founder of the fashion brand Yoox-Net-a-Porter. He is an innovative thought leader who has used digital technology to transform the way the fashion industry does business. According to the New York Times, Marchetti is regarded as the man who put fashion on the internet. He has used artificial intelligence and digital technology to transform the way people shop for luxury goods. He is also passionate about the importance of inclusive and responsible business models. Together, Marchetti and UNDP's Sharma will discuss the importance of using digital technology to achieve sustainable development. They will also talk about how it can put people and nature at the center of the fashion industry. Digital technology can help the fashion industry transform into an environment-friendly operation by allowing consumers and brands to satisfy their growing desire for sustainable fashion. It can also help prevent the fashion industry from becoming an outdated relic.

Digital transformation can help meet the challenge of reducing greenhouse gas emissions (Pūraitė et al., 2020). According to our analysis, the combined efforts of various industries could reduce global CO_2 emissions by around 26 billion metric tons from 2016 to 2025. If multiple sectors implement intelligent asset management and energy storage integration, it could reduce global CO_2 emissions by approximately 8 billion metric tons by 2025 (Shirazi & Falahat, 2012). In addition, the Global E-Sustainability Initiative (GeSI) believes that digital technologies can help reverse the high growth-high carbon trend. Through their framework, the organization found that the information and communications technology (ICT) sector can help users save 10 tons of CO_2 each year (Stratan, 2017).

Despite the potential of digital transformation, many challenges still need to be addressed to realize its full potential. One of these is the increasing amount of e-waste (Zos-Kior et al., 2020, Itce) many of the devices that are being discarded are not only hazardous to the environment but also have lost their value due to the lack of recycling options. A study conducted by the United Nations revealed that in 2014, around 40 million metric tons of e-waste were discarded globally. China and the U.S. were identified as the biggest producers of e-waste. The high-power consumption of data centers and the inefficient cooling systems they use contribute to the greenhouse gas emissions of these facilities. According to a report by the American Data Center Association, the electricity consumption by data centers is expected to increase by 20% by 2020.

Barriers to Achieving a Circular Economy by Digital Enterprises

Despite the potential of digital transformation, many challenges still need to be addressed to realize its full potential. One of these is the increasing amount of e-waste. A study conducted by the United Nations revealed that in 2014, around 40 million metric tons of e-waste were discarded globally. China and the U.S. were responsible for almost half of this amount. Using machine-to-machine and data analytics, they can identify and reduce the number of unused products and materials in their supply chains. The high-power consumption of data centers and the inefficient cooling systems they use contribute to the facility's greenhouse gas emissions. According to a report by the American Data Center Association, the electricity consumption by data centers is expected to increase by 20% by 2020. Despite the circular economy's potential, many barriers prevent it from being widely adopted. These include the complexity of the materials used, the geographical dispersion of supply chains, and the lack of recycling options.

Companies can reclaim valuable materials and components previously labeled as waste using end-of-life products. For instance, by recycling and reusing end-of-life products, Proctor & Gamble has created over $1 billion in value over the past five years. Extending the life of products can help companies reclaim the weight they lost due to their disposal. With various methods such as repairs, upgrades, and remanufacturing, they can keep their products economically advantageous for as long as they need to. The emergence of a sharing platform has created new business opportunities for consumers, companies, and micro-entrepreneurs. Through this concept, people can easily swap and borrow their unused goods. Through the concept of product as a service, retailers and manufacturers can start to realize the total cost of ownership of their products. For instance, Philips has started charging for the lighting used by its customers. Trust and adoption are two key factors consumers need to consider regarding digital technologies. According to a 2015 survey conducted by Accenture, almost half of consumers will have a connected device by 2020. This trend is expected to create tremendous opportunities for consumers to improve their energy efficiency and lower their utility bills. Connected devices must be built with the necessary security measures to protect their users' personal information. There are four ways to establish sustainability among digital enterprises such as.

a. The goal of implementing circular economy principles is to reduce the environmental impact of their operations and align them with their customers' goals. It can be done by launching products that can be reused or recycled and by offering services that help manage the use of existing assets.
b. Through digital technology, companies can improve the transparency of their supply chains and reduce their environmental impact. By benchmarking their suppliers and conducting in-house social responsibility assessments, they can identify areas where they can make the most impact.
c. Businesses should collaborate with other organizations to share unused capacity or waste streams and reduce their environmental impact. It can be done using platforms that allow them to optimize their routes and assets in the logistics industry.

d. The technology industry can also improve its reporting standards by establishing guidelines that cover the various aspects of environmental management. These guidelines should help companies identify and manage their e-waste and enhance the efficiency of their data centers.

Factors of Success for Digital Enterprises Toward Sustainability

Some critical factors may contribute to digital enterprises' overall success. Sustainability is the buzzword around which these factors revolve. The authors tried to explain these factors below.

Cost-effectiveness: The first step in creating a digital culture is cultivating a change mindset. This process can be done through small-scale experimentation and minimal financial resources. Content analytics tools can help you identify which content is most popular and which platforms are most utilized by your customers. An online community is an excellent way for people to connect and share their knowledge. Many organizations offer their members training through various platforms. Through digital support, an organization can reach out to its members at the right time and provide them with the right help. Machine learning and artificial intelligence can help improve the interactions between its members and the company.

Alternative digitalization process: Digitalizing pandemics has become more streamlined due to the rise of digital apps. One example would be using a content services platform (CSP), allowing participants to participate in the live action seamlessly. Before an organization can implement a content management system, it must first have a clear goal and plan. Doing so can help them become more digital-aware and successful.

Core digitalization: This chapter aims to discuss the various characteristics of a digital DNA that can be used to transform an organization. It also explores the multiple steps that can be taken to implement these traits. As technology changes the way work gets done, organizations must redesign their processes and activities to accommodate the new demands. It can be done with automation and human resources. Instead of following regulations, organizations can focus on their core business activities.

Digital leadership and workforce: Drive work through agile teams and have low-level hierarchies. A great leader will be able to maintain the same level of performance as before, but they will have to convince skeptical employees that their ideas are valuable. New talent categories should be more open and collaborative to support the evolution of the workforce. They should also develop the necessary skills to help the needs of their employees.

Digitalization experience and workforce: The journey's goal is to create an immersive work experience that combines the physical and virtual worlds. This process requires a combination of marketing and human resources to manage the evolution. The first step in developing a successful digital strategy is to identify the key factors that affect the development of the employees. It can be done through a design thinking approach that focuses on the customer and the employees. Having the right people in place will also help accelerate the digital transformation. The combination of marketing and human resources should be part of the strategy to manage the evolution. Human resources should be part of the strategy to enable the organization to thrive. It is also essential that the digital system includes consumer-grade experiences to help grow the organization. To do this, enterprises should start using analytics to identify the various factors influencing workforce development. The right digital system can help organizations accelerate their business by providing them with an insight-driven approach to help them navigate the multiple factors affecting their workforce.

Digitalized resources: The leaders committed to transforming their organizations' digital operations are committed to creating a collaborative and inclusive culture. It will allow them to develop and implement their digital strategy across all their departments. The complexity of digital transformation is immense. Due to the increasing number of new demands and threats. They should also plan their response in three phases. After the COVID-19 pandemic, technology leaders will transition into the next phase of their digital transformation. During this period, they will need to deploy unprecedented levels of collaboration and communication. Many will also have to make changes in their operations to prepare for the pandemic.

Findings and Discussion

Over half of the organizations surveyed have made significant changes to their strategy following the pandemic. Many of them have shifted to a more cloud-focused approach to operations. Over half of the respondents said they use automated process acceleration. After several years of being convinced of A.I.'s capabilities, many executives plan to implement these technologies in their organizations to improve their operations. While many of these goals focus on increasing the number of technical skills, the human element is still vital to success. The key to success is having the right people in the correct positions to handle the various changes brought about by digital technologies.

Despite the various obstacles that prevent technology-savvy executives from implementing effective strategies and methods to improve the efficiency of their operations, more than 75% of them still believe that the changes brought about by COVID-19 will have a positive impact on the way their customers behave. Despite the lack of a clear strategy for digital transformation, over 60% of the executives said they would use A.I. to improve the customer experience. Despite the positive impact of digital technologies, many executives still believe that their organizations are not yet ready to embrace cloud-based services fully.

Following the pandemic outbreak, many organizations started implementing a more cloud-based strategy. Almost 70% of them also said they would be increasing their digital transformation efforts. They also noted that they would be investing in blockchain, artificial intelligence, and the Internet of Things. The rise of automation and artificial intelligence has reduced the number of people needed to perform the tasks that humans previously handled—the stress caused by the pandemic affected the way executives planned for their company's future. It has caused them to lose their focus on improving the customer experience. Instead of developing a new vision, they focus on their current priorities.

Over the next two years, many executives expect their organizations to improve operational efficiency and cost management capabilities. Fifty-eight percent of executives said that new product development is their top priority, while 52% indicated that they would be focusing on new market entrants. Out of the digital enterprises considered for the study, almost 80% are focused on improving their operations. The majority of executives also believe that cash flow and liquidity are their top priorities. On the other hand, almost 70% of all organizations are focused on improving their cybersecurity. They also believe that the internet of things and cloud computing is the future of their businesses.

Various researchers also revealed that executives' priorities have shifted to include the multiple areas of crisis management and safety and security. They also noted that cost control is becoming more critical. The rise of platform-based businesses is also expected to significantly impact the supply-chain reliability of companies in the next couple of years. Despite the pandemic, the various sectors of the economy have different needs and want. The stock market's bifurcation evidences it. The healthcare industry was expected to be the most resilient sector following the pandemic outbreak. It was also predicted that home-based orders would increase. Although the pandemic was expected to hurt specific industries, such as manufacturing and financial services, leaders of enterprises believe that the number of collaborations and partnerships will increase over the next two years.

Despite the importance of flexibility, large enterprises have managed to operate successfully in volatile industries. They have been the ones that have held steady in this environment. Due to the importance of continuous improvement and the continuous pursuit of growth, organizations have started to reward their employees. It has led to a renewed focus on managing their change. To remain competitive, organizations need to re-examine their operations and plans for new ventures. This process should involve keeping in mind their goals and requirements and the procedures they follow to approve their resources.

During the outbreak, people avoided public places and used individual packaging. Fear of getting sick prompted many people to send their goods to their homes. Despite the episode, the public's interest in environmental issues has remained high. As a result, regulators became more concerned about the authenticity of products and brands. As a result, many companies have started implementing new regulations and standards.

Due to the increasing number of companies' interest in sustainable practices, many of them have started to consider improving the efficiency of their supply chains. Doing so will require implementing new processes and equipment. The coronavirus outbreak served as a wake-up call to the public about the increasing number of plausible and unexpected occurrences. It has also allowed companies to rethink their strategies and operations. The rapid emergence and evolution of cloud and digital transformation are expected to drive business competitiveness in the next couple of years, so executives need to step up their efforts to manage complexity. Unfortunately, the risks associated with the outbreak have prevented many people from returning to their daily lives. One of the most significant barriers to progress is the complexity of an organization's operations. According to studies, employee stress can trigger the rise of complexity. It can be attributed to the increasing number of tasks and responsibilities employees must perform to keep up with the market demands.

Although the outbreak was expected to harm specific industries, such as healthcare and financial services, it is also likely to impact others positively. Creating a thriving work environment requires the proper strategy, operations, and finance mix. It should also support people's mental health and well-being. Over the next couple of years, many companies are expected to expand their collaborations and partnerships.

Conclusion

There are three phases of a company's transformation: planning, journey, and lack of knowledge. Due to the emergence of the COVID-19 pandemic, various organizations have been affected. One of these is not implementing remote monitoring and video conferences. It has allowed people to change their attitudes. Unfortunately, COVID-19 didn't provide many long-term lessons. Instead, take advantage of the opportunity to talk to your team about what they need to prepare for in a crisis.

This chapter explores how technology can transform the way people work in the future. It involves identifying the best ways to use automation and human resources to improve efficiency. Due to the complexity of today's work environment, enterprises must adopt a cross-functional and agile approach to drive their business. This guide will help them identify the critical digital characteristics that will enable them to accelerate their digital transformation journey. The future of work is about establishing an open talent economy. It will allow organizations to build a more flexible workforce and improve their bottom-line performance. In an era of connected work, organizations are starting to use the power of experience design to improve their bottom-line results. The goal of today's enterprises is to create consumer-grade experiences that will enhance the productivity and engagement of their workforce. They will also introduce new talent categories such as private labels and freelance workers.

Before implementing a digital transformation strategy, organizations must identify the skills they need and the methods they can use to develop them. This process should also include developing a change management culture.

References

Abdallah, L. B. (2017). Towards an understanding of the Tunisian small and medium-sized enterprises' (SMEs) reality: New challenges and approaches. *International Journal of Technology Management & Sustainable Development, 16*(3), 205–207. https://doi.org/10.1386/tmsd.16.3.205_7

Agostini, L., Galati, F., & Gastaldi, L. (2019). The digitalization of the innovation process. *European Journal of Innovation Management, 23*(1), 1–12. https://doi.org/10.1108/ejim-11-2019-0330

Astafyeva, O. E. (2021). Sustainable development of industrial enterprises in the basis of a new form of economic activity organisation. *Vestnik Universiteta, 10*, 109–113. https://doi.org/10.26425/1816-4277-2021-10-109-113

Behera, M., Dash, M., Mohanty, A. K., & Das, S. (2019). Entrepreneurial ecosystem: A geographical upliftment. *Revista Espacios, 40*, 26–36. http://www.revistaespacios.com/a19v40n01/19400126.html

Das, S. (2020a). *Media impact of advertising on consumer buying behaviour a comparative study of different media*. Shiksha o Anusandhan University. http://hdl.handle.net/10603/273691

Das, S. (2020b). Innovations in digital banking service brand equity and millennial consumerism. *Digital transformation and innovative services for business and learning* (pp. 62–79). IGI Global. https://doi.org/10.4018/978-1-7998-5175-2.ch004

Das, S. (2021a). *Digital entertainment: The next evolution in service sector*. Springer Nature. https://doi.org/10.1007/978-981-15-9724-4. Print ISBN 978-981-15-9723-7. Online ISBN 978-981-15-9724-4

Das, S. (2021b). *Search engine optimization and marketing: A recipe for success in digital marketing* (1st ed.). CRC Press. https://doi.org/10.1201/9780429298509

Das, S. (2021c). A systematic study of integrated marketing communication and content management system for millennial consumers. In *Innovations in digital branding and content marketing* (pp. 91–112). IGI Global. https://doi.org/10.4018/978-1-7998-4420-4.ch005

Das, S., & Mondal, S. (2016). *Yours digitally: Patanjali*. Available at SSRN 2737299.

Das, S., Mondal, S. R., & Sandhu, K. (2022a). Music logos drive digital brands: An empirical analysis of consumers' perspective. *Journal of Strategic Marketing*, 1–16. https://doi.org/10.1080/0965254X.2022.2098526

Das, S., Mondal, S., Puri, V., & Vrana, V. (2022b). Structural review of relics tourism by text mining and machine learning. *Marketing, 8*(2), 25–34. https://doi.org/10.5281/zenodo.7358349

Duman, Ü. O., & Das, S. (2021). The dynamic rise of digital brands' market mavens in digital entertainment: A complete know how for curious customers. In *Digital entertainment* (pp. 147–172). Palgrave Macmillan. https://doi.org/10.1007/978-981-15-9724-4_8.

Duy, N. T., Mondal, S. R., Van, N. T. T., Dzung, P. T., Minh, D. X. H., & Das, S. (2020). A study on the role of Web 4.0 and 5.0 in the sustainable tourism ecosystem of Ho Chi Minh City, Vietnam. *Sustainability, 12*(17), 7140. https://doi.org/10.3390/su12177140

Gupta, D. K., Jena, D., Samantaray, A. K., & Das, S. (2019). HRD climate in selected public sector banks in India. *Revista ESPACIOS, 40*(11).

Irawan, A. (2020). Challenges and opportunities for small and medium enterprises in Eastern Indonesia in facing the COVID-19 pandemic and the new normal era. *TIJAB (The International Journal of Applied Business), 4*(2), 79. https://doi.org/10.20473/tijab.v4.i2.2020.79-89

Ivanov, I., Orlova, L., & Vlasova, T. (2021). Digitalization as a factor of sustainable development of industrial enterprises. *IOP Conference Series: Earth and Environmental Science, 666*(6), 062047. https://doi.org/10.1088/1755-1315/666/6/062047

Jain, S., Jain V., & Das, S. (2018). Relationship analysis between emotional intelligence and service quality with special evidences from Indian banking sector. *Revista ESPACIOS, 39*(33).

Laksmanawati, J., & Yuniawan, A. (2021). Women and the digitalization strategies of micro, small, and medium enterprises in the new normal era. *Petra International Journal of Business Studies, 4*(1), 55–64. https://doi.org/10.9744/ijbs.4.1.55-64

Minh-Nhat, H. O., Majerova, J., & Das, S. (2022a). Taking steps to help the phoenix rise from ashes: A roadmap for digital enterprises to develop new sustainable goals. In *Sustainable development and innovation of digital enterprises for living with COVID-19* (pp. 35–51). Springer. https://doi.org/10.1007/978-981-19-2173-5_3

Minh-Nhat, H. O., Nguyen, H. L., & Mondal, S. R. (2022b). Digital transformation for new sustainable goals with human element for digital service enterprises: An exploration of factors. In *Sustainable development and innovation of digital enterprises for living with COVID-19* (pp. 85–103). Springer. https://doi.org/10.1007/978-981-19-2173-5_6

Mohanty, P.C., Dash, M., Dash, M., & Das, S. (2019). A study on factors influencing training effectiveness. *Revista Espacios, 40*, 7–15. http://www.revistaespacios.com/a19v40n02/19400207.html

Mondal, S. R. (2020a). *Factors influencing store image loyalty and satisfaction from customer perspective an empirical study in retail sector*. Shiksha o Anusandhan University. http://hdl.handle.net/10603/273688

Mondal, S. R. (2020b). A systematic study for digital innovation in management education: An integrated approach towards problem-based learning in Vietnam. In *Digital innovations for customer engagement, management, and organizational improvement* (pp. 104–120). IGI Global. https://doi.org/10.4018/978-1-7998-5171-4.ch006

Mondal, S. R. (2021). A systematic study of new age consumer engagement and exploration for digital entertainment for over-the-top platforms in various digital media. In *Innovations in digital branding and content marketing* (pp. 113–133). IGI Global. https://doi.org/10.4018/978-1-7998-4420-4.ch006

Mondal, S. R., & Das, S. (2021). Examining diabetic subjects on their correlation with TTH and CAD: A statistical approach on exploratory results. In *Machine learning and the internet of medical things in healthcare* (pp. 153–177). Academic Press. https://doi.org/10.1016/B978-0-12-821229-5.00007-0

Mondal, S. R., Das, S., Musunuru, K., & Dash, M. (2017). Study on the factors affecting customer purchase activity in retail stores by confirmatory factor analysis. *Revista ESPACIOS, 38*(61).

Mondal, S. R., Majerova, J., & Das, S. (2022). Sustainable development and innovation of digital enterprises for living with COVID-19. Springer. https://doi.org/10.1007/978-981-19-2173-5

Mondal, S. R., Virgilio, F. D., & Das, S. (2022). *H.R. analytics and digital H.R. practices*. Springer. https://doi.org/10.1007/978-981-16-7099-2

Nadanyiova, M., & Das, S. (2020). Millennials as a target segment of socially responsible communication within the business strategy. *Littera Scripta, 13*(1), 119–134. https://doi.org/10.36708/Littera_Scripta2020/1/8

Ngoc-Vinh, N., Tien-Dung, P., & Das, S. (2022). Digitization of Business; Need for Recombination During COVID-19. In *Sustainable development and innovation of digital enterprises for living with COVID-19* (pp. 1–17). Springer. https://doi.org/10.1007/978-981-19-2173-5_1

Nguyen, T. D. M., Mondal, S. R., & Das, S. (2022). Digital entrepreneurial transformation (det) powered by new normal sustainable developmental goals (n-SDGs): Elixir for growth of country's economy. In *Sustainable development and innovation of digital enterprises for living with COVID-19* (pp. 69–84). Springer. https://doi.org/10.1007/978-981-19-2173-5_5

Pūraitė, A., Adamonienė, R., & ŽEmeckė, A. (2020). Sustainable digitalization in public institutions: Challenges for human rights. *European Journal of Sustainable Development, 9*(3), 91. https://doi.org/10.14207/ejsd.2020.v9n3p91

Ravi, S., & Mondal, S. R. (2021). Digital entertainment based do it yourself content & advertisement as a factor of driving force for trust & customer patronage. In *Digital entertainment* (pp. 127–146). Palgrave Macmillan. https://doi.org/10.1007/978-981-15-9724-4_7

Sharma, E., & Das, S. (2020). Measuring impact of Indian ports on environment and effectiveness of remedial measures towards environmental pollution. *International Journal of Environment and Waste Management, 25*(3), 356–380. https://doi.org/10.1504/IJEWM.2019.10021787

Sharma, E., & Das, S. (2021). Integrated model for women empowerment in rural India. *Journal of International Development*, 1–18.https://doi.org/10.1002/jid.3539

Sharma, E., Nigam, N., & Das, S. (2020). Measuring gap in expected and perceived quality of ICT enabled customer services: A systematic study of top ten retailers of India. *International Journal of Applied Systemic Studies, 9*(2), 159–184. https://doi.org/10.1504/ijass.2020.113260

Shirazi, M., & Falahat, S. (2012). Compact urban form, question or solution? Examining the compact city in the Middle Eastern context: Challenges and opportunities. *International Journal of Urban Sustainable Development, 4*(2), 246–259. https://doi.org/10.1080/19463138.2012.694817

Singh, L. B., Mondal, S. R., & Das, S. (2020). Human resource practices & their observed significance for Indian SMEs. *Revista ESPACIOS, 41*(07). http://www.revistaespacios.com/a20v41n07/20410715.html

Singh, S., & Das, S. (2018). Impact of post-merger and acquisition activities on the financial performance of banks: A study of Indian private sector and public sector banks. *Revista Espacios Magazine, 39*(26), 25.

Siri, R., & Das, S. (2021). A study on processing of information storage & use of new age consumers in digital wellness sector through story telling & creating interest. In *Digital entertainment* (pp. 45–63). Palgrave Macmillan. https://doi.org/10.1007/978-981-15-9724-4_3

Siri, R., Mondal, S. R., & Das, S. (2020). Hydropower: A renewable energy resource for sustainability in terms of climate change and environmental protection. In *The handbook of environmental chemistry*. Springer. https://doi.org/10.1007/698_2020_635

Stratan, D. (2017). Success factors of sustainable social enterprises through circular economy perspective. *Visegrad Journal on Bioeconomy and Sustainable Development, 6*(1), 17–23. https://doi.org/10.1515/vjbsd-2017-0003

Tien-Dung, P. H. A. M., Majerova, J., & Das, S. (2022a). Phases of possible recovery of digital enterprises in new normal business for living with COVID-19 times: Opportunities for a new era in sustainable development goals. In *Sustainable development and innovation of digital enterprises for living with COVID-19* (pp. 19–33). Springer. https://doi.org/10.1007/978-981-19-2173-5_2

Tien-Dung, P., Nguyen, X. D., & Das, S. (2022b). Sustainable goal achievement by digital revolution during and after pandemic; how much one wins and losses: A bird's eye view for future planning. In *Sustainable development and innovation of digital enterprises for living with COVID-19* (pp. 133–147). Springer. https://doi.org/10.1007/978-981-19-2173-5_8

Tiwari, M., & Mondal, S. R. (2022). Technological dynamism of workforce management for effective education service delivery during and after Covid-19. In *H.R. analytics and digital H.R. practices* (pp. 199–215). Palgrave Macmillan. https://doi.org/10.1007/978-981-16-7099-2_8

Van, N. T. T., Vrana, V., Duy, N. T., Minh, D. X. H., Dzung, P. T., Mondal, S. R., & Das, S. (2020). The role of human-machine interactive devices for post-COVID-19 innovative sustainable tourism in Ho Chi Minh City, Vietnam. *Sustainability, 12*(22), 9523. https://doi.org/10.3390/su12229523

Yegen, C., & Mondal, S. R. (2021). Sharenting: A new paradigm of digital entertainment of new age parenting and social media. In *Digital entertainment* (pp. 213–231). Palgrave Macmillan. https://doi.org/10.1007/978-981-15-9724-4_11

Zos-Kior, V., Ilin, V., Kyryliuk, I., & Solod, O. (2020). Digitalization in realization of ecological and economic principles of managing sustainable development of Agrarian enterprises. *Mechanism of an Economic Regulation, 2*, 29–36. https://doi.org/10.21272/mer.2020.88.02

PART IV

New Normal Digitalization and Human Interventions

CHAPTER 10

Defining the New Digitalized Normal: Implications of A.I., IoT, and Blockchain for Business Growth

Subhra R. Mondal and Subhankar Das

Introduction

Using artificial intelligence (A.I.), enterprises could develop effective strategies to fight against COVID-19 (Mondal et al., 2022). This report explores the various ethical and human rights concerns associated with the applications of A.I. This section aims to provide the public with an overview of the different new technologies used in the fight against COVID-19 (Tiwari & Mondal, 2022). There is no consensus on which type of applications were used. The recovery of the global economy must not be rushed back to pre-COVID-19 levels (Mondal & Das, 2021).

S. R. Mondal · S. Das (✉)
The Honors Programme, Southern Star Institute of Management, Duy Tan University, Da Nang, Vietnam
e-mail: subhankardas@duytan.edu.vn

S. R. Mondal
e-mail: subhraranimondal@duytan.edu.vn

S. R. Mondal et al. (eds.), *New Normal in Digital Enterprises*,
https://doi.org/10.1007/978-981-19-8618-5_10

Doing so could cause severe environmental damage. It is also essential to plan for the recovery of the various sectors of the economy. Getting people back to work can help prevent further economic damage (Sharma & Das, 2021). However, it is also essential that the recovery plan includes measures to help people live a more sustainable life (Das, 2021a).

A well-designed recovery plan can help address the various issues related to climate change and sustainable development. It can stimulate the uptake of new technologies and improve the efficiency of the electricity distribution. The pandemic that hit the world in 2019 and early 2020 accelerated society's digital transformation. Many people started working from home and attending classes remotely (Siri & Das, 2021). During the pandemic, despite the digital transformation's positive effects, it has highlighted the various security concerns that remain (Ravi & Mondal, 2021).

Although the pandemic highlighted the potential of digital technology in improving traditional medicine, it is unlikely that society will fully embrace it (Duman & Das, 2021). As the world becomes increasingly dependent on digital technology, its proper use must be ensured. It is also essential that enterprises create comprehensive plans to help bridge the digital divide (Yegen & Mondal, 2021). This chapter aims to discuss the various steps that governments are taking to implement digital strategies following the pandemic.

The use of technology during COVID-19 has increased due to the pandemic. It has also led to new ways to engage with one another and improve digital accessibility. Although some people may choose not to use tech, it is still expected that this will positively impact e-commerce and health (Das, 2021b). The rise of COVID-19 has raised the pressure on firms to provide high-quality connectivity. It could also create new digital segments globally (Sharma et al., 2020). Despite the increase in usage, the share of people aged 50–74 who use the Internet daily remained low compared to the average age of daily Internet users (Mondal, 2020a). Despite the increasing number of people using the Internet, skills gaps stay in various countries. One of these is the digital gender divide, which is an issue that needs to be addressed globally (Das, 2020a). Before COVID-19, the e-commerce market was growing globally (Das, 2021c). However, during the pandemic, the gap between large and small enterprises operating in this space has widened (Mondal, 2021). It has caused the diffusion of digital technology.

Despite the rise of big data, it is still a desirable scenario for various sectors and countries (Duy et al., 2020; Van et al., 2020). The government's priority is to build a sustainable future through digital transformation. Many countries have already started implementing strategies and programs related to this. Over 60 countries will have national A.I. strategies by 2020, and several countries have begun developing 5G strategies (Sharma & Das, 2020; Singh et al., 2020; Siri et al., 2020). The virtuous circle between innovation and digital transformation is also a significant issue for many countries (Das, 2020b; Mondal, 2020b; Nadanyiova & Das, 2020). Despite the potential risks associated with the use of digital technology, many countries still believe that its use should not be restricted to commercial interests (Behera et al., 2019; Gupta et al., 2019; Mohanty et al., 2019). Despite the positive effects of digital transformation, it is still not enough to ensure a more resilient and inclusive future. The COVID-19 pandemic has reinforced the need for a coordinated approach to digital transformation (Jain et al., 2018; Mondal et al., 2017; Singh & Das, 2018). Therefore, a framework has been created to help governments implement policies that will help shape a more inclusive digital future. Here, the authors tried to give various framework policies as follows;

a. Digital connectivity has become an integral part of businesses and schools as it allows them to access and manage their data.
b. As the number of people and firms using digital technology increases, governments need to step up their efforts to prepare them for the digital economy.
c. Digital innovation is a critical component of digital transformation. It can help governments achieve efficiency gains and stimulate new business models.
d. Following the emergence of digital tools following the pandemic, more attention needs to be given to ensuring that the public and businesses trust the digital environment.
e. The rise of digital technology has raised concerns about the future of the workforce. It also led to an increase in people working from home. As the recovery from the pandemic continues, policymakers will need to look into the various regulations and labor market structures that apply to the industry.
f. As more people spend their time online, governments should also step up their efforts to support the well-being of their citizens.

g. The rise of tech companies has also raised concerns about the impact of market fragmentation on startups and small businesses. As the number of online services continues to decrease, governments should also consider the effect of this trend on society.

A strong and robust economy can help develop a sustainable digital economy. After the global pandemic affected various regions, governments implemented emergency measures to support recovery (Das & Mondal, 2016). As the outbreak started to subside, they focused on preparing stimulus plans to boost the economy. This chapter aims to analyze the various steps governments have taken to support recovery. It also explores the various factors that can help improve long-term stability. This chapter focuses on the implementation of sustainable development projects following the COVID-19 guidelines. This section will discuss the various issues that will affect this process.

COVID-19 has highlighted various threats to the economy and society. It has also led to the rapid spreading of the pandemic. The global economic crisis highlighted the need for focusing on short-term growth over long-term resilience. The loss of over 300 million jobs during the global financial crisis highlighted the widening gap between the haves and have-nots (Ngoc-Vinh et al., 2022). Despite the situation, public awareness about the various frailties has increased. The global economy's vulnerability is most alarming due to the multiple threats that climate change poses. Multiple factors such as air pollution and biodiversity loss have already affected the development and operations of the global economy (Tien-Dung et al., 2022a).

Aside from the economic crisis, various natural disasters have also affected regions like the Philippines and Bangladesh. Without urgent action on climate change, greenhouse gas emissions will increase and threaten the global economy. The recovery from the global financial crisis is expected to be very short-lived (Nguyen et al., 2022). The interrelated environmental issues related to climate change and biodiversity loss could have catastrophic effects on society. Without a long-term recovery, businesses and individuals will feel the impact of the crisis on their well-being. As the government and enterprises implement stimulus measures, they must take a step back and reflect on the causes of the problem (Minh-Nhat et al., 2022a).

Despite the positive signs that emerged during the recovery, many recovery plans have fallen short. In 2020, G20 finance ministers committed to supporting a sustainable recovery. However, a survey conducted in 2020 revealed that most people still prioritize environmental issues (Minh-Nhat et al., 2022b).. Climate change became a political issue in 2019 after millions of people participated in peaceful protests. The World Economic Forum also highlighted the importance of addressing biodiversity loss and climate change in its report in 2020. Despite the political climate change and the growing concerns about climate change, the World Economic Forum still believes that a sustainable recovery is the best way to address these issues (Tien-Dung et al., 2022b).

Revival of Enterprises by Digital Innovations

Building back better refers to the recovery from COVID-19, which focuses on improving communities' resilience and reducing the costs of future disasters. The challenge of building a new global economy following a disaster is different than that of a physical disaster. Despite the economic recovery, the risks of a more sustainable recovery remain (Mondal et al., 2022). Despite the global efforts to prevent disasters, the actions taken during the crisis can still help minimize the risks. Digital enterprises have the following objectives.

a. To build a more sustainable recovery, enterprises should consider implementing recovery measures designed to provide immediate benefits and have long-term goals (Das et al., 2022a). The recovery should be focused on improving the conditions of society and addressing the various factors that affect people's well-being. A stimulus package should also be created to support the environment (Das et al, 2022b). Doing so can expose potential overlap between the policies of different stimulus plans.
b. Policymakers should consider introducing progressive taxation and subsidies.

Digital enterprises of different countries may follow the below-described steps for achieving these pertinent objectives.

a. Climate change is a critical issue that the U.S. and other countries must address to avoid experiencing the worst effects of the global economic crisis. Supporting recovery measures that reduce greenhouse gas emissions is also essential to prevent the worst effects of climate change.

b. Climate change is also a critical issue that businesses need to address to stimulate recovery efforts. Climate change is expected to have a significant impact on society through the development of infrastructure networks. To minimize the effects of climate change, new infrastructure projects should implement effective risk management strategies. However, it is essential that these projects are resilient and do not increase their vulnerability.
c. The protection of biodiversity is an integral part of human health and the global economy. However, the increasing destruction of forests and land-use changes are threatening the stability of these ecosystems. Stimulus packages should include the value of biodiversity in their decisions. Also, governments should take measures to improve the management of their biodiversity support. Doing so can help improve the resilience of ecosystems.
d. Fostering innovation is also a critical component of achieving sustainable development goals. It helps develop new processes and improve the efficiency of existing ones. The recent emergence of the COVID-19 pandemic has highlighted the need for stimulus packages focused on addressing cultural norms and behavior change. One of the most effective ways to reduce greenhouse gas emissions is by increasing the use of public transport. It can help improve the cleanliness of streets and reduce crowding.
e. The emergence of the pandemic has also highlighted the need for more effective policies and procedures to support the sustainable development of global supply chains. Although the exact environmental impact of the shift is not yet clear, policies must be put in place to help improve the performance of local supply chains.

Digital Enterprise and Financial Investment for Sustainable Practices

Financial investment is paramount for sustainability. There are some prime focal points where digital enterprises can have an economic stimulus package.

a. The proposed stimulus package could help accelerate the transition to an electricity system resilient to climate change. However, it should also consider the various changes needed to make the system more energy efficient.
b. The low oil and gas prices are expected to harm the renewable energy sector. It will reduce the amount of money invested in the renewable energy industry.
c. The proposed stimulus package could also help decrease greenhouse gas emissions. It can also help communities by increasing the availability of electricity.
d. One of the most critical components of a green recovery package is energy efficiency. It can help improve the resilience of the electricity system and deliver various benefits. In the U.S. and Europe, it is estimated that over 3.3 million individuals participate in the energy efficiency industry.
e. Beyond reducing greenhouse gas emissions, improving the energy efficiency of buildings is also a vital component of the country's climate change strategy. It can help lower the cost of energy. However, it is not yet clear if the electric vehicle program will positively impact greenhouse gas emissions.
f. The challenges that small-scale projects face when scaling up energy conservation and distributed energy resources during the recovery are often related to their liquidity constraints. Therefore, governments should use their existing resources to create a pipeline of projects that can be quickly replicated.
g. Aside from infrastructure projects, the stimulus package should also include training programs to improve the skills of individuals in the energy sector. It could involve the deployment of smart grids and electric storage technologies.

The housing crisis highlighted various issues related to the sector. It has become more visible due to the increasing number of situations where people struggle to live comfortably. Buildings are responsible for over 30% of global emissions. Buildings are often built to last. However, they use energy inefficiently, leading to costly repairs and maintenance. Despite the benefits of energy efficiency, many barriers remain. The various factors that can prevent the rapid deployment of energy efficiency programs prevent countries from making the most of their resources. Different government policies have been enacted to promote energy efficiency.

These include tax breaks for home appliances and the scrapping of inefficient ones. The rise of the COVID-19 pandemic could also cause people to move away from more densely populated areas and into areas with lower infection risk. It could also contribute to the increasing demand for less dense housing. Urban renewal strategies can help mitigate the effects of climate change by encouraging people to live in areas they want to. They can also help cities improve their resilience to climate change.

Smart Accessibility and Mobility Among Different Stakeholders

The government should stimulate the passenger transport sector by providing long-term support to the car industry. However, it should also ensure that these measures are linked to the necessary environmental improvements. While supporting the car industry, governments also need to encourage the development of systems that will allow people with limited mobility to use their cars. One example is the lock-in system, which prevents people from leasing low-emission vehicles.

Despite the increasing number of people working in the car industry, public transport is still an essential part of the transportation system. However, it also faces new challenges, such as the increasing number of reluctant people to use public transportation. Cities such as Reno and Los Angeles have already seen a drop in traffic due to implementing public transport projects. They should also encourage collaboration between public transportation agencies and businesses to improve the availability of transportation.

As the recovery process begins, it's time for cities to develop new transportation options that will create more jobs and improve air quality. Over a hundred cities in the U.S. have already implemented emergency measures to create more active transportation. Doing so could help make the transition to a more sustainable transportation system by reducing air pollution and improving the efficiency of implementing policies related to the reduction of fossil fuel subsidies. Despite the increasing number of people working in the car industry, public transport is still an essential part of the transportation system. However, it also faces new challenges, such as the increasing number of people who are reluctant to use public transport.

Digital Enterprises and Their Contribution to the Circular Economy

The complexity of the global value chains highlighted the importance of improving resilience. However, implementing new strategies and procedures could inadvertently increase the environmental impacts of their operations. The transportation and production of raw materials and finished goods are significant contributors to global greenhouse gas emissions. Due to the increasing number of materials being produced and shipped, the rate of recycling is low.

The complexity of supply chains can also create risks of disruption. It can also contribute to the concentration of upstream actors. The reduction of greenhouse gas emissions was beneficial for developed nations. The complexity of supply chains can also make it hard to evaluate potential disruption risks. If a company plans to reduce its supply chains, it should avoid increasing its environmental impact. The recovery measures implemented by governments can help reduce greenhouse gas emissions and improve the efficiency of the global supply chains. One of the most effective ways to reduce waste is using circular value chains. This process involves the diversion of waste from landfills into reprocessing and recycling.

The circular value chain can also help reduce greenhouse gas emissions. It can allow companies to shift their consumption away from primary raw materials. An increase in digital technology can also help companies minimize their risks and improve resilience. It can also help governments identify opportunities to reduce greenhouse gas emissions. Large companies can also benefit from circular value chains by reducing their administrative burden.

On the other hand, the implementation of digital technology can help improve the efficiency of industrial processes. Automation can also help companies reduce their labor costs. However, as the technology becomes more prevalent, it will require more active management to maximize its benefits.

New Standard Solutions for Digital Enterprises

Digital transformation has become a common concept in recent years as businesses look to improve their operations and sustain themselves. They also want to minimize their expenses. Due to the rapid emergence of digital technologies, many companies have started exploring new ways of maintaining themselves. One of these is the New Normal Solutions, which are strategic and digital solutions that help organizations adapt to the changes brought on by digital transformation. The New Normal Solutions framework helps enterprises implement digital transformation strategies to help them achieve their goals. It also provides a set of solutions that can be used to improve the operations of their businesses. The authors of the New Normal Solutions framework help organizations identify their specific challenges and develop practical digital transformation strategies.

The New Normal solutions help organizations achieve their goals and improve their efficiency. They can be used to integrate various activities within their value chain. Many companies have started implementing digital solutions to protect their customers and employees from the COVID-19 pandemic. The rapid deployment of these tools will continue during the recovery period. When developing a New Normal strategy, it is crucial to consider the mindsets of the various stakeholder groups that will be affected by the changes. These include the expectations of the business, the need for continuous improvement, and the deployment of advanced analytics. Five years ago, digital adoption wasn't widely acknowledged. Today, it is regarded as a mainstream technology used by mainstream businesses and consumers. Banks are now moving toward remote sales and service teams. Schools are also starting to focus on digital learning, and grocery stores are now turning to online ordering and delivery as their primary business models. In addition, schools are also starting to use digital classrooms. Doctors are also delivering telemedicine services. After the pandemic, businesses worldwide are facing the challenge of returning to a more stable environment.

The rise of digital services is one of the main structural changes affecting the way businesses operate. According to a survey, over 75% of people who have never used a digital channel said they would continue doing so once the situation returns to normal. The rapid emergence and evolution of digital media will severely affect the operations and capital of digital laggards. Due to the situation's complexity, many companies will rely on historic data and forecasts to make critical decisions. New data will help predict future trends and provide insight into companies' operations. It will allow them to make better decisions and improve their efficiency. The rise of remote work has significantly impacted the way businesses perform their jobs. It has allowed them to respond faster to their customers' inquiries.

Recovery by the Digital Model

Despite the various changes in the digital world, many companies still lack the necessary resources and capabilities to realize their potential fully. Many companies are already migrating to a digital recovery model, which allows them to retain and hire talent from anywhere. This model also enables them to launch new initiatives such as artificial intelligence. Therefore, implementing a digital recovery plan is essential. It focuses on four key areas: improving operational efficiency, accelerating digital investments, leveraging new data, and developing technology capabilities.

As companies accelerate their digital-first transformation, they see the emergence of e-commerce as a core component of their operations. It has delivered robust growth and improved its basket size. However, it is still crucial for them to consider other factors such as safety and customer experience. Through the availability of various services and solutions, automobile companies can now offer their customers a wider variety of products and services. In response, airlines are also developing new ways to improve the travel experience.

Implementation by Artificial Intelligence and Other Techniques

Every day, hundreds of decisions are made in an airline's operations. These decisions include the routes that the company should operate, the number of crew members it should have, and the number of meals it should order. Due to the pandemic outbreak, many of these decisions will need to be re-implemented. After the financial crisis of 2008, many companies had to re-build their economic models. These models will need to be updated for various data sets, such as time series and unemployment data. Developing and implementing new predictive models requires collaboration between multiple departments and organizations. For instance, a car-parts supplier had to create a forecasting model to identify potential issues with its supplies. The goal of a sophisticated model is to inform a supplier about the problems that it should address. It can also help the company identify potential solutions. Aside from sales, a model can also help other business areas.

Before developing new models, executives should first identify the areas they should focus on and develop. It will allow them to implement the changes immediately. Machine-learning techniques have permitted organizations to create new analytical models capable of handling large data sets. The complexity of a program's development process can be reduced by investing in new digital solutions and freeing up resources. It can be done using software development tools and the technology stack. These models can help organizations reduce their overall cost of operations. For instance, reducing non-critical jobs can also improve their profitability.

A cloud-based platform and a software development pipeline are the most critical features of today's technology environments. They can be quickly implemented and help improve the speed of development. Before developing a plan, the executives must identify the appropriate cost structure and the providers that can deliver the best value. Having the right technology stack modernization is also vital for companies. Instead of focusing on the traditional hiring and retaining talent issues, executives should start implementing strategies that will help their teams deliver better software. It can be done with the right tools and processes.

Collaborative Efforts for Digitalization

The current economic crisis has caused organizations to rethink their operations and how they work with their customers. Many companies have already started implementing agile methods, but they are not yet ready to scale them to the speed needed to respond to the crisis's changes. Due to the situation's complexity, it can take a long time to develop a new product or service. Many companies have limited resources and cannot scale up fast enough to handle the complexity. So here, this optimum utilization is given below.

a. One of the most effective ways to speed up the digital delivery process is to establish a digital factory. This type of facility allows companies to build and scale their digital infrastructure.
b. To implement remote work models, identify the areas that could provide the most productivity. Then, design the new models and operationalize them in three months.
c. To become successful in digital-led recovery, leaders must first reset their digital agendas and develop a strategy that will meet the needs

of their customers. They must also measure their progress against these goals.

Effective Post-Pandemic Digital Transformation

In today's digital world, do traditional roles in information technology encourage innovation? The authors explored this issue by examining the role of transformational leaders in helping organizations innovate. They also found that the lack of formal IT governance practices can prevent them from doing so. The authors found that stable cultures have a more moderate effect on innovation than change cultures. It suggests that traditional roles in information technology can still encourage innovation. This chapter then presents policies that support digital transformation. The coronavirus pandemic has highlighted the need for more public services and institutions globally.

The coronavirus outbreak highlighted the importance of regional development models that put the interests of citizens at the center. Digital disruption can help address this issue by transforming industries and creating new competitive conditions. The coronavirus outbreak has also strengthened the case for digital transformation. Countries can improve their services and build a more credible system through digital tools. This is because they can use advanced technologies such as artificial Intelligence (A.I.), machine learning, and blockchain to transform their economies. A comprehensive public policy effort is needed to align various national development goals with digital agendas. This would help promote the adoption of digital transformation policies.

The coronavirus pandemic has severely affected the economies of various countries. It is expected to harm the global economy in the next two years. Some of the effects of the pandemic include a drop in commodity prices and a decline in tourism. The pandemic has also affected the most vulnerable groups, self-employed individuals. Before the outbreak, about 40% of the workforce was not covered by a safety net. Since the crisis started, various measures have been implemented to support these individuals. Multiple measures have been taken to support the recovery efforts, such as providing cash transfers and credit guarantees. However, these actions are not enough to address the global economic constraints. Aside from these, other fiscal measures are also needed to support the recovery efforts. These include improving public debt management. As a result, countries in the region need policies to address their structural issues. The rapid emergence and evolution of digital technologies have brought various challenges and opportunities. While the government is making efforts to seize these, it is also essential that policies are implemented to create a more conducive environment for the growth of digital economies.

Aside from a sectoral approach, policymakers also need to step up their efforts to support the digital transformation. It can be done by developing instruments and procedures that allow cross-border transactions. The gap between the productivity of developed and developing countries has remained high over the years. Most countries' competitiveness is attributed to their abundance of natural resources and low-skilled labor. Due to the lack of technological innovation in developed nations, their productivity has decreased. It is mainly due to the country's outdated production and technical infrastructure. Despite the various advantages of digital technology, many businesses still face challenges in adopting it. It is because digital entrepreneurship is very challenging.

Although the potential impact of digital technology on productivity is not automatic, various factors can affect it. One of the most critical factors that businesses need to consider when implementing digital transformation is the availability of skilled labor. This can be done through the training and education of various professions and roles. One of the most effective ways that digital transformation can help improve the efficiency of the transport networks is with e-commerce platforms. It can help boost the competitiveness of the logistics industry. The digital transformation could also create productivity gaps due to the varying characteristics of small and medium-sized enterprises.

Despite the technological advancements over the years, the gap between small and large businesses remains significant. It is due to the lack of digital technology adoption among SMEs. Despite the current economic crisis, digital transformation is still expected to play a vital part in the recovery. Robotic systems and other technological innovations could do this. However, it is still essential to have the proper legal framework to support the development of the digital economy. There is a considerable potential for R&D in the digital economy to produce new business models and innovations. It can be done through policies that support this space.

The digital transformation can help households improve their lives and employment opportunities (Taskinsoy, 2019). It can also make them more self-sufficient. The digital revolution can also help workers get more work done by making their jobs more automated. The rise of the digital course can also help people work more efficiently and safely (Treiblmaier & Tumasjan, 2022). The coronavirus outbreak highlighted the importance of a balanced work–life balance. It also made workplaces safer. Unfortunately, many people can still not fully capitalize on the digital tools that are available to them. It can lead to long-term losses and social issues (Tyagi et al., 2020). Despite the rise of the digital revolution, the percentage of households that use the Internet remains higher in the wealthiest quintile than in the poorest quintile (Zia, 2021). The rapid emergence and evolution of new job opportunities are crucial factors for workers. Many people are reluctant to improve their skills due to the threat of automation. Many low-skilled workers can also not take advantage of the training opportunities that are available to them. It is also a concern for informal workers.

Digital Transformation and Governance

The rise of the middle class and the public's expectations have raised the need for better services. Despite the progress made in sharing governance, institutions are still not responding adequately to the changes brought about by digital transformation. These policies aim to help public institutions improve their operations and administration by developing new procedures and rules to help them manage the digital transformation (Bouncken et al., 2019). A regulatory framework that helps protect consumers is also vital to maintaining the public's trust and confidence in the economy. It can also help prevent security breaches. The governing of the digital economy is a critical issue that governments should address. (Chang, 2022) Due to the nature of the data collected and stored in the healthcare industry, it can be very vulnerable to unauthorized access and theft. This issue has become a significant concern for the government. Artificial intelligence (A.I.) apps are commonly based on international models. The increasing number of these has raised ethical concerns (Lim, 2021). Data collected on the algorithms used for these apps can be biased or inaccurate, affecting the integrity of their work.

Through social media, citizens can help strengthen their trust in government. During a crisis, gathering public opinion can be beneficial. It can lead to corruption. One way to prevent this is by establishing a central purchasing body that can help citizens track their financial transactions (Nitsche & Straube, 2021). The rise of digital technology has allowed governments to improve their public governance through tools such as the Internet. This ecosystem is composed of various government agencies and non-government organizations. Through digital transformation, governments can lower their costs and provide better customer services by improving the efficiency of their operations (Priya & Revathy, 2021). They can also increase the number of people involved in their decision-making processes. The education sector is one of the fastest-growing areas in terms of digital technologies. Using data, governments can improve their services and develop effective public policies. During the coronavirus pandemic, which affected several countries, emergency medical services reduced the number of people sent to hospitals (Sandner et al., 2020). Many countries have started using smartphones to collect data about the disease's spread. Authorities can monitor the space and evolution of coronaviruses. This method can help them identify areas where the disease might be spreading.

Conclusion

The poor and marginalized communities disproportionately feel the burden of pandemics. As governments respond to this new outbreak, they need to build a comprehensive public health response. Despite the availability of resources, countries can still not guarantee the sustainability of their social protection systems. Although the pandemic can trigger modifying their plans, they can still be strengthened. The digital economy can help address the COVID-19 crisis by contributing to the development of resilient communities. In response to the outbreak, various support organizations have stepped up their efforts to help startup ecosystems. Developing nations can benefit from international collaboration to build their digital infrastructures and improve communication capabilities. Through regional cooperation initiatives, countries can create standard regulations and procedures that will help them develop their digital strategies and promote their voice on international platforms. The European Union's Digital Agenda shows how this can be done. Developing nations can also benefit from international cooperation to overcome the various challenges associated with the digital economy.

References

Behera, M., Dash, M., Mohanty, A. K., & Das, S. (2019). Entrepreneurial ecosystem: A geographical upliftment. *Revista Espacios, 40*, 26–36. http://www.revistaespacios.com/a19v40n01/19400126.html

Bouncken, R. B., Kraus, S., & Roig-Tierno, N. (2019). Knowledge- and innovation-based business models for future growth: Digitalized business models and portfolio considerations. *Review of Managerial Science, 15*(1), 1–14. https://doi.org/10.1007/s11846-019-00366-z

Chang, H. (2022). Business communication digitalized: Accelerated transformations under the new normal. *Business Communication Research and Practice, 5*(1), 1–3. https://doi.org/10.22682/bcrp.2022.5.1.1

Das, S. (2020a). *Media impact of advertising on consumer buying behaviour a comparative study of different media*. Shiksha o Anusandhan University. http://hdl.handle.net/10603/273691

Das, S. (2020b). Innovations in digital banking service brand equity and millennial consumerism. In *Digital transformation and innovative services for business and learning* (pp. 62–79). IGI Global. https://doi.org/10.4018/978-1-7998-5175-2.ch004

Das, S. (2021a). *Digital Entertainment: The Next Evolution in Service Sector*. Springer Nature. https://doi.org/10.1007/978-981-15-9724-4. Print ISBN978-981-15-9723-7. Online ISBN978-981-15-9724-4

Das, S. (2021b). *Search engine optimization and marketing: A recipe for success in digital marketing* (1st ed.). CRC Press. https://doi.org/10.1201/9780429298509

Das, S. (2021c). A systematic study of integrated marketing communication and content management system for millennial consumers. In *Innovations in digital branding and content marketing* (pp. 91–112). IGI Global. https://doi.org/10.4018/978-1-7998-4420-4.ch005

Das, S., & Mondal, S. (2016). *Yours digitally: Patanjali*. Available at SSRN 2737299.

Das, S., Mondal, S. R., & Sandhu, K. (2022a). Music logos drive digital brands: An empirical analysis of consumers' perspective. *Journal of Strategic Marketing*, 1–16. https://doi.org/10.1080/0965254X.2022.2098526

Das, S., Mondal, S., Puri, V., & Vrana, V. (2022b). Structural review of relics tourism by text mining and machine learning. *Marketing, 8*(2), 25–34. https://doi.org/10.5281/zenodo.7358349

Duman, Ü. O., & Das, S. (2021). The dynamic rise of digital brands' market mavens in digital entertainment: A complete know how for curious customers. In *Digital entertainment* (pp. 147–172). Palgrave Macmillan. https://doi.org/10.1007/978-981-15-9724-4_8

Duy, N. T., Mondal, S. R., Van, N. T. T., Dzung, P. T., Minh, D. X. H., & Das, S. (2020). A study on the role of Web 4.0 and 5.0 in the sustainable tourism ecosystem of Ho Chi Minh City, Vietnam. *Sustainability, 12*(17), 7140. https://doi.org/10.3390/su12177140

Gupta, D. K., Jena, D., Samantaray, A. K., & Das, S. (2019). HRD climate in selected public sector banks in India. *Revista ESPACIOS, 40*(11).

Jain, S., Jain V., & Das, S. (2018). Relationship analysis between emotional intelligence and service quality with special evidences from Indian banking sector. *Revista ESPACIOS, 39*(33).

Lim, J. (2021). How blockchain and IoT-based decentralized system solves ULD rental problem of airlines. *Archives of Business Research, 9*(10). https://doi.org/10.14738/abr.910.11164

Minh-Nhat, H. O., Majerova, J., & Das, S. (2022a). Taking steps to help the phoenix rise from ashes: A roadmap for digital enterprises to develop new sustainable goals. In *Sustainable development and innovation of digital enterprises for living with COVID-19* (pp. 35–51). Springer. https://doi.org/10.1007/978-981-19-2173-5_3

Minh-Nhat, H. O., Nguyen, H. L., & Mondal, S. R. (2022b). Digital transformation for new sustainable goals with human element for digital service enterprises: An exploration of factors. In *Sustainable development and innovation of digital enterprises for living with COVID-19* (pp. 85–103). Springer. https://doi.org/10.1007/978-981-19-2173-5_6

Mohanty, P. C., Dash, M., Dash, M., & Das, S. (2019). A study on factors influencing training effectiveness. *Revista Espacios, 40*, 7–15. http://www.revistaespacios.com/a19v40n02/19400207.html

Mondal, S. R. (2020a). *Factors influencing store image loyalty and satisfaction from customer perspective an empirical study in retail sector*. Shiksha o Anusandhan University. http://hdl.handle.net/10603/273688

Mondal, S. R. (2020b). A systematic study for digital innovation in management education: An integrated approach towards problem-based learning in Vietnam. In *Digital innovations for customer engagement, management, and organizational improvement* (pp. 104–120). IGI Global. https://doi.org/10.4018/978-1-7998-5171-4.ch006

Mondal, S. R. (2021). A systematic study of new age consumer engagement and exploration for digital entertainment for over-the-top platforms in various digital media. In *Innovations in digital branding and content marketing* (pp. 113–133). IGI Global. https://doi.org/10.4018/978-1-7998-4420-4.ch006

Mondal, S. R., & Das, S. (2021). Examining diabetic subjects on their correlation with TTH and CAD: A statistical approach on exploratory results. In *Machine learning and the internet of medical things in healthcare* (pp. 153–177). Academic Press. https://doi.org/10.1016/B978-0-12-821229-5.00007-0

Mondal, S. R., Das, S., Musunuru, K., & Dash, M. (2017). Study on the factors affecting customer purchase activity in retail stores by confirmatory factor analysis. *Revista ESPACIOS, 38*(61).

Mondal, S. R., Majerova, J., & Das, S. (2022). Sustainable development and innovation of digital enterprises for living with COVID-19. Springer. https://doi.org/10.1007/978-981-19-2173-5

Mondal, S. R., Virgilio, F. D., & Das, S. (2022). HR analytics and digital HR practices. *Springer*. https://doi.org/10.1007/978-981-16-7099-2

Nadanyiova, M., & Das, S. (2020). Millennials as a target segment of socially responsible communication within the business strategy. *Littera Scripta, 13*(1), 119–134. https://doi.org/10.36708/Littera_Scripta2020/1/8

Ngoc-Vinh, N., Tien-Dung, P., & Das, S. (2022). Digitization of Business: Need for Recombination During COVID-19. In *Sustainable development and innovation of digital enterprises for living with COVID-19* (pp. 1–17). Springer. https://doi.org/10.1007/978-981-19-2173-5_1

Nguyen, T. D. M., Mondal, S. R., & Das, S. (2022). Digital entrepreneurial transformation (det) powered by new normal sustainable developmental goals (n-SDGs): Elixir for growth of country's economy. In *Sustainable development and innovation of digital enterprises for living with COVID-19* (pp. 69–84). Springer. https://doi.org/10.1007/978-981-19-2173-5_5

Nitsche, B., & Straube, F. (2021). Defining the "new normal" in international logistics networks: Lessons learned and implications of the COVID-19 pandemic. *WiSt—Wirtschaftswissenschaftliches Studium, 50*(11), 16–25. https://doi.org/10.15358/0340-1650-2021-11-16

Priya, S. S., & Revathy, S. (2021). Security enhanced collaborative business model approach using blockchain and IoT. *International Journal of Business Innovation and Research, 1*(1), 1. https://doi.org/10.1504/ijbir.2021.10041340

Ravi, S., & Mondal, S. R. (2021). Digital entertainment based do it yourself content & advertisement as a factor of driving force for trust & customer patronage. In *Digital entertainment* (pp. 127–146). Palgrave Macmillan. https://doi.org/10.1007/978-981-15-9724-4_7

Sandner, P., Gross, J., & Richter, R. (2020). Convergence of blockchain, IoT, and AI. *Frontiers in Blockchain, 3*. https://doi.org/10.3389/fbloc.2020.522600

Sharma, E., & Das, S. (2020). Measuring impact of Indian ports on environment and effectiveness of remedial measures towards environmental pollution. *International Journal of Environment and Waste Management, 25*(3), 356–380. https://doi.org/10.1504/IJEWM.2019.10021787

Sharma, E., & Das, S. (2021). Integrated model for women empowerment in rural India. *Journal of International Development*, 1–18. https://doi.org/10.1002/jid.3539

Sharma, E., Nigam, N., & Das, S. (2020). Measuring gap in expected and perceived quality of ICT enabled customer services: A systematic study of top ten retailers of India. *International Journal of Applied Systemic Studies, 9*(2), 159–184. https://doi.org/10.1504/ijass.2020.113260

Singh, S., & Das, S. (2018). Impact of post-merger and acquisition activities on the financial performance of banks: A study of Indian private sector and public sector banks. *Revista Espacios Magazine, 39*(26), 25.

Singh, L. B., Mondal, S. R., & Das, S. (2020). Human resource practices & their observed significance for Indian SMEs. *Revista ESPACIOS*, *41*(07). http://www.revistaespacios.com/a20v41n07/20410715.html

Siri, R., & Das, S. (2021). A study on processing of information storage & use of new age consumers in digital wellness sector through story telling & creating interest. In *Digital entertainment* (pp. 45–63). Palgrave Macmillan. https://doi.org/10.1007/978-981-15-9724-4_3

Siri, R., Mondal, S. R., & Das, S. (2020) Hydropower: A renewable energy resource for sustainability in terms of climate change and environmental protection. In *The handbook of environmental chemistry*. Springer. https://doi.org/10.1007/698_2020_635

Taskinsoy, J. (2019). Blockchain: Moving beyond bitcoin into a digitalized world. *SSRN Electronic Journal*. https://doi.org/10.2139/ssrn.3471413

Tien-Dung, P. H. A. M., Majerova, J., & Das, S. (2022a). Phases of possible recovery of digital enterprises in new normal business for living with COVID-19 Times: Opportunities for a new era in sustainable development Goals. In *Sustainable development and innovation of digital enterprises for living with COVID-19* (pp. 19–33). Springer. https://doi.org/10.1007/978-981-19-2173-5_2

Tien-Dung, P., Nguyen, X. D., & Das, S. (2022b). Sustainable goal achievement by digital revolution during and after pandemic; how much one wins and losses: A bird's eye view for future planning. In *Sustainable development and innovation of digital enterprises for living with COVID-19* (pp. 133–147). Springer. https://doi.org/10.1007/978-981-19-2173-5_8

Tiwari, M., & Mondal, S. R. (2022). Technological dynamism of workforce management for effective education service delivery during and after Covid-19. In *HR analytics and digital HR practices* (pp. 199–215). Palgrave Macmillan. https://doi.org/10.1007/978-981-16-7099-2_8

Treiblmaier, H., & Tumasjan, A. (2022). Editorial: Economic and business implications of blockchain technology. *Frontiers in Blockchain*, *5*.https://doi.org/10.3389/fbloc.2022.857247

Tyagi, N., Gautam, S., Bharadwaj, J., & Goel, A. (2020). B-IoT (Blockchain Internet of Things): A way to enhance IoT applications via blockchain. *SSRN Electronic Journal*. https://doi.org/10.2139/ssrn.3747476

Van, N. T. T., Vrana, V., Duy, N. T., Minh, D. X. H., Dzung, P. T., Mondal, S. R., & Das, S. (2020). The role of human-machine interactive devices for post-COVID-19 innovative sustainable tourism in Ho Chi Minh City, Vietnam. *Sustainability, 12*(22), 9523. https://doi.org/10.3390/su12229523

Yegen, C., & Mondal, S. R. (2021). Sharenting: A new paradigm of digital entertainment of new age parenting and social media. In *Digital entertainment* (pp. 213–231). Palgrave Macmillan. https://doi.org/10.1007/978-981-15-9724-4_11

Zia, M. (2021). B-DRIVE: A blockchain based distributed IoT network for smart urban transportation. *Blockchain: Research and Applications, 2*(4), 100033. https://doi.org/10.1016/j.bcra.2021.100033

CHAPTER 11

Digital Revolution Complements Sustainable Goal Achievement During and After Pandemic: A Research Agenda

Smriti Tandon and Subhankar Das

Introduction

The digital revolution, which includes virtual reality, 3D printing, and artificial intelligence, has entered the public discourse in many countries (Adner, 2016). The digital revolution is becoming a key driver of society's transformation. Potential to transform society must be considered along with the various goals and initiatives of the 2030 Agenda (Amit & Zott, 2015). The digital revolution will transform society's operation and how it thinks about sustainability. It is also a driver of disruptive change.

S. Tandon
Graphic Era Deemed to be University, Dehradun, India
e-mail: smriti.tandon.mba@geu.ac.in

S. Das (✉)
The Honors Programme, Southern Star Institute of Management, Duy Tan University, Da Nang, Vietnam
e-mail: subhankardas@duytan.edu.vn

S. R. Mondal et al. (eds.), *New Normal in Digital Enterprises*,
https://doi.org/10.1007/978-981-19-8618-5_11

The digital revolution is the key enabler of sustainable development. However, it can create many negative externalities (Anjum et al., 2017). The digital process is a powerful force that can propel change through its transformative capabilities. At the same time, it can also carry solid social disruptive power if handled with caution and innovation (Benner & Tripsas, 2010). The report highlights the various benefits of the digital revolution and its potential negative impacts. It also warns about the challenges of addressing the digital divide, which excludes many people currently living in poverty (Besharov & Smith, 2014). The chapter also outlines the various conditions that a successful digital transformation needs to be successful. It also talks about the social implications of the digital revolution and the related governance considerations. Humanity's dilemma is to live within a safe and just Earth system. A growing list of global risks threatens our planet's stability. While preserving the planet's resiliency depends on optimizing socioeconomic development, it also requires improving human welfare.

In 2015, the world's prominent organizations adopted the 2030 Agenda for sustainable development. The 2030 Agenda is a far-reaching and time-bound plan that aims to achieve ambitious goals (Candau & Gbandi, 2019). It was developed through a comprehensive consultation among civil society and national governments. It includes a set of goals and targets that can be achieved by 2030. It also has plans for social and economic development. The rise of the digital revolution, which includes artificial intelligence (A.I.), virtual reality, and additive manufacturing, has been widely debated.

The digital transformation will have a transformative effect on society and economies. It will also change the way we view and act on sustainable development. Below are some of the critical considerations that should be considered when exploring the potential of the digital revolution for sustainable development. Digitalization aspects are discussed in the following elements (Chellappa & Mukherjee, 2021). Here authors tried to give a holistic discussion of the six factors of the digital revolution.

A. ***Digitalization & Society***: The digital revolution, which started in the 1950s, is a third civilizational revolution accelerating. It marks the emergence of new technical systems that can enhance human cognition. In 2019, the real numbers were spontaneously generated in a deep neural network trained to recognize visual objects (Bruzzone, 2021). The performance revealed the characteristics of the animal and human number discriminations. Digital Anthropocene is a newly configured context that needs to be rethought and developed to address the various challenges and opportunities (Cropanzano et al., 2017). Digital technologies can help us achieve a Sustainable Anthropocene by accelerating the decarbonization of all sectors and by contributing to a reduction of greenhouse gas emissions (Garipağaoğlu, 2021). This report and many others show that digital platforms can help achieve these goals. Although digital media can help accelerate the decarbonization of various sectors, this transition will not be an automatic process (Foss & Saebi, 2016). Technological revolutions have led to resource and greenhouse-gas-intensive growth patterns for many years. To avoid disrupting the current trends, a radical reversal of these trends is needed to create pathways toward sustainable development (Huang & Rust, 2018). The report identifies six fundamental transformations required to achieve the 17 sustainable development goals.

B. ***Digital automation for SDG***: Digital technologies can help accelerate the decarbonization of the world, and they can also contribute to a more sustainable Anthropocene. According to the TWI2050 report, digital technologies can help accelerate the decarbonization of the world, and they can also contribute to a more sustainable environment (Joung et al., 2021). Although digital technology can help accelerate the decarbonization of the world, it will not be an automatic process. Until now, the transition to a more sustainable future has been primarily triggered by the effects of the previous industrial revolutions (Kaldeli et al., 2016). A radical change in current trends helps achieve the sustainable development goals (SDGs) and maintain a long-term sustainable future. The report TWI2050 identifies six fundamental transformations needed to reach the goals.

The 2030 Agenda for Sustainable Development is a set of goals that should be regarded as mid-point indicators toward achieving sustainable development by 2050 (Lekan et al., 2020). There is an urgent need for policies and incentives to support this goal. The Six Essential Mechanisms can be linked to enable the Six Fundamental Challenges of Sustainable Development using digital technologies (Kim et al., 2014). These include mobilizing market forces and environmental policies that support sustainable development. Using digital technology to transform planning and markets processes for a sustainable future is an essential component of this strategy (Korkeamäki & Kohtamäki, 2020). It can help pave the way for more effective and efficient use of resources and a better understanding of goals and milestones for various sectors.

C. ***Effect of digitalization on society and sustainable development goals***: The rapid emergence and evolution of digital technologies have raised the concerns of society's leaders (Nishiura & Mimura, 2021). If not appropriately addressed, these changes could threaten community cohesion and undermine the 17 Sustainable Development Agenda. The four significant challenges facing humanity are inequality, education systems, labor market, and dissipative forces (Marinova et al., 2016). There is also a potential rise of political power due to the so-called Big Five companies (e.g., Amazon, Facebook, Google, and Microsoft). With the potential to improve governance and steering policies, digitalization could also help address the four slippery slopes.

D. ***Digital technology and effect on humans; Quantum leap beyond current trend***: The next decade will see profound changes in society and economies because of digital transformation. It requires policymakers and organizations to step up their efforts to understand digital change's various effects and prepare for the inevitable structural changes (Ng & Vargo, 2018). With the rise of digital disruption, we are entering a new era of cultural and economic realities. We can now improve our cognitive capabilities to make informed decisions using artificial intelligence and virtual reality. The emergence of autonomous systems based on artificial intelligence will fundamentally transform society (Ojo, 2022). By processing vast amounts of data, AI-based machines will change how we consume and work (Rust & Huang, 2014). They will also enable new treatments for diseases and provide personalized forecasts for individuals. The evolution of human intelligence has had no rival since humans first appeared. With the rise of artificial intelligence, it is now considered a primary task for humanity to meet sustainable development goals (Sick et al., 2019). There are many challenges in developing and governing sustainable pathways aligned with the profound societal transitions that are happening right now. How can our legal systems keep up with the rapid technological change brought about by the rise of virtual and self-learning environments? How can automation help people become more prosperous while leaving no one behind? What ethical guardrails should be placed in place to transform humans and artificial intelligence? What role should human judgment play in the development of machine learning? Can we avoid making systems that can control human behavior? These questions are not part of the 2030 Agenda. They illustrate the complexity of the tasks that need to be done to build a sustainable global society. These tipping points are based on the various characteristics of our communities and consider the multiple configurations of society that will shape up over the next decade (Turber et al., 2014). Climate change and its impacts are triggering various tipping points on Earth's planetary boundaries due to the digital growth patterns not diverted toward sustainability (Urbinati et al., 2021). These points are also triggered by the erosion of civil society due to the increasing use of digital surveillance and authoritarian rule. Others are concerned about the uncontrolled human enhancement methods that will transform humanity.

E. ***Circular economy with human intervention and the digital revolution***: The paradox of the digital Anthropocene is that although the digital process has created unprecedented opportunities for transformative change, it is also endangering the very foundations of sustainable development. The evolution of the circular economy is already within reach (Vargo & Lusch, 2017). It could allow us to decouple wealth creation from the consumption of resources. The fusion of knowledge breakthroughs and technological innovations can create new opportunities for humanity. This potential is evidenced by new biomimetics, synthetic biology, and composite materials. The digital dynamics could enable new cultural and institutional innovation (Westerlund et al., 2014). It could help build a global society based on shared values and create new common goods perspectives. Virtual worlds could help people better understand cultural diversity. In parallel, digitalized voting procedures could help democratize democracy and improve local governance efficiency. The convergence of digital technologies and physical and cognitive capabilities will enhance human performance (Wirtz et al., 2010). Human life spans have increased dramatically due to the achievements of scientists and athletes in the field of knowledge and health. The rise of virtual reality and the fusion of various facts could lose privacy and control over one's personal data and personality. It could threaten the freedom of people in the future. The loss of control is considered one of the biggest threats to the human Anthropocene. Despite the hype about the potential of digital technology, machines controlling humans is a real danger. Despite the uncertainties of the future, the direction of change remains unknown.

Therefore, the importance of steering change must be the highest priority. The digital revolution presents a unique opportunity to build responsible knowledge societies that act toward sustainability (Xu et al., 2014). The potential risks posed by the various technologies and artificial intelligence that can be used in the digital era can be tackled if the two communities can come together. To achieve sustainable development, we will need to connect the tremendous potential of the twenty-first century's technological breakthroughs with the most challenging tasks. The current era of digitalization presents various challenges in addressing the multiple issues of the Digital Anthropocene. How can we interact with A.I., and how can we maintain democracy in the face of technological change? In times of digital disruption, authoritarian governments and powerful business actors pose significant threats to democracy and civil rights.

F. ***Societal challenges for the digital revolution***: Technology can face many societal challenges, such as increasing digitalization. However, it is also clear that the benefits of technology can be maximized for a sustainable future. The potential for significant progress in education, health, and wealth creation is undisputed. There are also societal impacts brought about by our changes in how we live and work. With so many changes happening in society, governments and organizations are at a critical crossroads (Zott & Amit, 2010). The crucial question is whether these changes can be managed or regulated to be accommodated efficiently. Unfortunately, most of the consequences are only acknowledged after the fact when introducing new technologies. The digital revolution can help inform the public about the many advantages of new technologies (Das & Mondal, 2016). However, it can also restrict their full potential due to public concerns. The following principles help create an interconnected system architecture that will support digital and sustainable transformation management.

a. Education and science are the key factors that people need to understand and shape the digital shifts. Through education, people will shape the future of digital transformation. Science will also play a crucial role in transforming knowledge networks.
b. Digitalized development needs to improve the capabilities of public institutions to manage and understand the digital world.
c. Experimental spaces are the key to cultivating a culture of learning and experimenting.
d. The digital revolution will have global impacts, such as the modernization of the U.N.
e. The 2030 Agenda can be seen as a new social contract that will guide how we think about the world beyond 2030. It will involve new development models focused on the environment and society.

The six transformations are actions that can help us achieve a better life for ourselves and our planet. They are linked to power dynamics that can affect the development of humans. The various processes that make up the six transformations are managed according to their evolution. If they are appropriately managed, they can serve as an entry point for achieving the goals of the SDGs.

Sustainable Development and Digitalization: Prerequisites

The sustainable development goals are a set of goals that aim to improve the lives of all people on Earth. These goals are sometimes referred to as the triple bottom line. They are focused on human, environmental, and economic development. The 2030 Agenda emphasizes the importance of good governance. The 17 SDGs are universal and referred to as holistic goals. They are designed to provide a globally agreed framework for achieving goals and management directions (Mondal et al., 2017). The U.N.'s sustainable development goals are a unique concept that sets a new social contract for the world. They are designed to address the most critical issues of our times. Since human activity has already breached planetary safe spaces, all countries must now shoulder their responsibilities to achieve the goals of the sustainable development agenda (Jain et al.,

2018). Some prerequisites are needed for sustainable development, which is described as follows.

a. ***Development mindset for sustainable growth***:
The digital revolution has the potential to improve the well-being of people globally. However, its potential to do so is overshadowed by the uncertainties associated with the nature and future of work. Despite the significant productivity growth that technology has provided, there is no guarantee that it will create more jobs (Das et al., 2022b). Understanding the economic conditions that left those without a college education is pressing for policymakers (Singh & Das, 2018). There is no immediate solution to the issue of productivity growth, which most countries are likely to face. We need to move away from a planned obsolescence model that involves the disposal of non-productive resources. Instead, it should be a circular economy that encourages the reuse of resources.
Unfortunately, despite the rapid emergence of digital technologies, most do not yet have the proper incentive to reuse finished products and materials (Das et al., 2022a). It is mainly due to the increasing number of features in devices. Due to the growing number of marketing campaigns, consumers are now buying replacement products that are often only as good as the new ones (Gupta et al., 2019). Although they may be obsolete, many new products still have the same functionality. In the U.K., it has been estimated that around four times as many mobile phones are still unused as those currently used. Over 30% of household appliances are still in working condition in Germany (Behera et al., 2019). The growth of green industries through the circular economy can help in an increasing number of jobs created by the digital technologies (Ngoc-Vinh et al., 2022).
b. ***Social inclusions and Societal development:***
Digital technologies have made their way across much of the world, but they also have characteristics that make them more prone to social inequality (Tien-Dung et al., 2022a). The digital divide shows that the various benefits of using digital technologies have fallen short of being inclusive (Minh-Nhat et al., 2022a, b). This divide is often relevant to different social groups and regions across the globe. Equalities are affected by various factors such as access to digital technologies, education, and affordability (Mohanty et al.,

2019). They also impact the economy through job losses, technological breakthroughs, and increased productivity. The promise of digital inclusion is only realized if it reaches the currently left behind people.

There may be warning signs of the dangers of algorithms and their potential to leave people behind. Digitalization can help bridge the digital divide, but it can also cause more problems if not managed well. Most of the barriers to digital inclusion are related to access (Nguyen et et al., 2022). For instance, around half of the world's population still lacks Internet access. This list does not consider the various factors that affect the quality of service, such as reliability and affordability (Nadanyiova & Das, 2020). Most of the time, digital technologies are more prone to diffuse than some basic ones. Internet access can also bridge the divide between people living in different countries. A 2016 study by the World Bank showed that economic disparities remained large even across generations (Minh-Nhat et al., 2022a, b). Unfortunately, there are still barriers to accessing these resources, such as the lack of electricity and Internet costs. Open-source databases such as Wikipedia help ensure the spread of information globally. Due to the steady growth of Internet access and electricity, the proportion of people with Internet access is expected to reach almost full coverage by 2030 (Mondal, 2020b). Through social media, groups and movements can become mainstream. However, this platform can also reduce the diversity of perspectives, creating echo chambers. The development of echo chambers could disrupt a well-informed public debate (Tien-Dung et al., 2022b). It could also lead to the rise of mental disorders (Das, 2020b). Internet access can help spread values and norms according to human rights principles. These can be spread through various means, such as social media and television, to enable people to make informed decisions about their bodies and sexual and reproductive health.

c. ***Green Sustainability:***

The Anthropocene is characterized by the rapid emergence of new technologies and their potential to improve society's well-being. However, achieving a sustainable future requires addressing the various factors that affect it, such as land use and climate change (Mondal et al., 2022). Technology's increasing adoption has led to the depletion of Earth's natural resources and climate change. The increasing popularity of digital devices, especially smartphones, has raised concerns about their lack of Earth's resources (Sharma & Das, 2020). In addition, their use increases the amount of e-waste. The number of mobile devices has more than doubled in less than three decades, and this rapid increase in consumption places additional stress on the environment (Singh et al., 2020). A report released in 2018, for instance, indicated that the greenhouse gas emissions from digital technology will increase by around 2% in 2020 and 8% in 2025. A reduction in energy and carbon footprint can be achieved with new, highly efficient technologies. However, this trade-off can be appropriately managed (Siri et al., 2020). Unfortunately, the pace of the digital revolution has caused regulators to adopt policies that are often enacted after the fact. It leads to the development of too short regulations and tends to impose harsh conditions on the use of new technologies.

d. ***E-governance and digitalization boost for sustainability:***
Peace and good governance are critical to sustainable human development. The digital transformation can also shape the political actors' roles in shaping this ideal. The digital transformation presents new political and security challenges, such as the rise of cyberwars and drone wars. Although states and societies can now rely on digital tools to address these challenges, they still need to make the most of them. These reforms are still the same as those proposed in the TWI2050 framework. The digital transformation can also create new opportunities for governments and communities to work together to achieve the SDGs (Van et al., 2020). Getting the most out of the digital transformation involves ensuring that technology is not used in contested governance spaces. It is because it can magnify existing inequalities. Despite the potential benefits of the digital economy, there are still many ways to use it for good. Strictly following the rules and regulations can still be very beneficial for humanity. One of these challenges is having someone with deep technical knowledge evaluate multiple platforms (Duy et al., 2020). After transitioning to a free and open-source platform, government agencies must rethink their procurement process. There are also opportunities for governments to implement flexible digital media.

e. ***Digital revolution and healthcare***
The use of technologies for healthcare has always been an essential part of the system. In particular, advancements in medical imaging

have led to a reduction in invasive procedures (Mondal, 2021). Due to the rapid emergence of digital technology, healthcare has become more accessible and personal. For instance, ultrasound technology is now available via smartphones. Various new technologies such as virtual reality and mobile health have emerged in the past couple of years. With the rapid emergence and acceptance of fitness trackers, people can now monitor their health in real-time. Through these tools, people can monitor various aspects of their health, such as their weight, heart rate, and sugar levels. It has led to an increase in their motivation to stay healthy. New mobile technologies have also saved many lives. According to researchers, the availability of telemedicine could help improve the healthcare system in developing countries. For instance, a remote community in Australia was able to establish a self-service clinic that provided patients with access to regular medicines (Das, 2021c). Telemedicine is expected to play a massive role in the future, as it can help overcome the shortage of skilled health professionals and reduce travel times for patients.

Patients can receive regular updates from their doctors (Das, 2020a). They can also keep track of their vital statistics using their smartphones. Doctors can also implant wearable devices into their bodies to collect data on various health conditions. About 90% of prostate surgeries are carried out using robots (Mondal, 2020a). Due to their ability to perform the surgery, they are being used more frequently. The ability to perform virtual surgery allows surgeons to improve their capabilities and provide an alternative to traditional open surgery. And it's also being used to treat mental disorders (Sharma et al., 2020).

Discussions

This chapter aims to describe the impact of intelligent technologies on business models. Instead of focusing on the effects of the Internet of Everything on a single firm, we used a service ecosystem perspective to examine how intelligent things can affect resource flows in a network. Using all the digitally published and printed secondary studies, we observed that the intelligent technologies' impact on business models increases with smartness (Das, 2021b). Less sophisticated smart objects may also potentially affect business models at the micro-level. The authors then draw on our analysis to develop propositions about intelligent devices and their impact on value creation. The various propositions authors observed can be used as starting points for scholars who want to create new insights into innovative technologies' impact on business models.

Micro-Level Policies

Ending things with smartness may transform the way we deliver services and products to our customers. This benefit can improve the efficiency of our workforce and provide better convenience. According to our observation, high-level smartness can be autonomous and work seamlessly with other people. These tools help customers save time and effort. For instance, when a car's software update is needed, they can ask the customer to arrange for it without the need for the customer to interact with a car maintenance firm (Yegen & Mondal, 2021). The increasing level of smartness helps customers accept intelligent objects. However, this level of acceptance can be affected by various factors such as security and privacy (Duman & Das, 2021). Unfortunately, exchanging data between different bright things can expose sensitive information such as passwords and tell hackers. It could expose them to various security risks.

At high levels of smartness, customers might feel apprehensive about their data being shared with other parties. It could affect their trust in the businesses that use the data collected by these connected objects. Prior studies suggest that companies should communicate the types of data they collect and the purposes for which it is used (Ravi & Mondal, 2021). Despite the advantages of having a variety of intelligent objects, businesses have a hard time developing effective business models that are flexible enough to handle the customers' demands (Siri & Das, 2021). Concerns about the security of the data collected by intelligent objects and the companies that use it can also decrease the acceptance of these products.

From a frontline employee's perspective, intelligent technology tools that allow people to do their jobs better are valuable. This expectation aligns with the literature that shows that adopting technology leads to better customer service and increased employee productivity. However, achieving this advantage can be challenging since implementing innovative technology often involves a lot of work (Das, 2021a). For instance, implementing innovative technology can require much time and effort in healthcare. The right tools and training can help people become more productive and leave their desk-bound colleagues to handle their tasks more smoothly (Sharma & Das, 2021). However, with high levels of smartness, people might feel that they lose their autonomy and decision-making power.

Meso-Level Policies

Business models are not static, which helps to understand how intelligent digital technologies work. They are constantly being enacted through the actors' resource integration practices. Data flows are the backbone of digitalized business models. They dictate how value is created and how business models can create conflicts or complementarity (Mondal & Das, 2021). Human capabilities can help minimize risks and make sure that value creation occurs. It is because they can manage the complexity of the process and provide the necessary tools and resources to support value creation (Tiwari & Mondal, 2022). For instance, if a home security system is connected to a mobile phone and a doorbell, this could result in a business model that is not acceptable to insurance companies. Humans play a dual role in the digitalized ecosystem, as they are both moderators and instigators of conflicts among connected business models. They both have their own needs to support value creation.

Macro-Level Policies

As part of our discussion, we consider the various challenges of studying digital technologies and intelligent things. This chapter shows how connected things can transform the value ecosystems of industries. This convergence leads to new service ecosystems and new business models. Industries do it in the form of digitalized technology-enabled services. One of the most challenging factors in establishing value co-creating ecosystems is choosing the right partners. It is an area where we often encounter conflicts (Mondal et al., 2022). Despite the various transitions in the industry, many firms still struggle to identify the opportunities and challenges presented by intelligent device-enabled industry convergence. For instance, smart cars are becoming more connected, creating intra-industry connectivity between various actors within the car industry, such as parking companies, car-sharing firms, and petrol stations.

Conclusion

The concept of smartness refers to the ability of devices to adapt to new situations and act autonomously. This enables them to perform tasks that are not possible with traditional methods. The concept of mutual benefit-sharing agreements is often used to describe the various arrangements that will be created because of the digital revolution. It is widely believed that these will involve new institutional arrangements. The centralization of the platform industry has the potential to reduce competition and increase social inequality. However, this process also creates various trade-offs that could lead to a fragmented technical landscape. The development of intelligent technologies can only be realized if all parties can share data. Therefore, it is expected that new institutional arrangements will be created to manage the power of monopolies. The way people use their products and services will influence the development of intelligent technologies. As a result, the interactions between various parties will become more sophisticated. This chapter aims to discuss the various levels of smartness, their implications for sustainable development, and business models. Through case studies, researchers can gain a deeper understanding of these concepts. The rise of smartness is expected to have a wide-ranging impact on various stakeholder groups. Therefore, it is important that researchers understand its various effects.

References

Adner, R. (2016). Ecosystem as structure. *Journal of Management, 43*(1), 39–58. https://doi.org/10.1177/0149206316678451

Amit, R., & Zott, C. (2015). Crafting business architecture: The antecedents of business model design. *Strategic Entrepreneurship Journal, 9*(4), 331–350. https://doi.org/10.1002/sej.1200

Anjum, A., Sporny, M., & Sill, A. (2017). Blockchain standards for compliance and trust. *IEEE Cloud Computing, 4*(4), 84–90. https://doi.org/10.1109/mcc.2017.3791019

Behera, M., Dash, M., Mohanty, A. K., & Das, S. (2019). Entrepreneurial ecosystem: A geographical upliftment. *Revista Espacios, 40*, 26–36. http://www.revistaespacios.com/a19v40n01/19400126.html

Benner, M. J., & Tripsas, M. (2010). The influence of prior industry affiliation on framing in nascent industries: The evolution of digital cameras. *SSRN Electronic Journal*. https://doi.org/10.2139/ssrn.1721005

Besharov, M. L., & Smith, W. K. (2014). Multiple institutional logics in organizations: Explaining their varied nature and implications. *Academy of Management Review, 39*(3), 364–381. https://doi.org/10.5465/amr.2011.0431

Bruzzone, S. (2021). A posthumanist research agenda on sustainable and responsible management education after the pandemic. *Journal of Global Responsibility, 13*(1), 56–71. https://doi.org/10.1108/jgr-05-2021-0045

Candau, F., & Gbandi, T. (2019). Trade and institutions: Explaining urban giants. *Journal of Institutional Economics, 15*(6), 1017–1035. https://doi.org/10.1017/s1744137419000328

Chellappa, R. K., & Mukherjee, R. (2021). Platform preannouncement strategies: The strategic role of information in two-sided markets competition. *Management Science, 67*(3), 1527–1545. https://doi.org/10.1287/mnsc.2020.3606

Cropanzano, R., Anthony, E. L., Daniels, S. R., & Hall, A. V. (2017). Social exchange theory: A critical review with theoretical remedies. *Academy of Management Annals, 11*(1), 479–516. https://doi.org/10.5465/annals.2015.0099

Das, S. (2020a). *Media impact of advertising on consumer buying behaviour a comparative study of different media*. Shiksha o Anusandhan University. http://hdl.handle.net/10603/273691

Das, S. (2020b). Innovations in digital banking service brand equity and millennial consumerism. In *Digital transformation and innovative services for business and learning* (pp. 62–79). IGI Global. https://doi.org/10.4018/978-1-7998-5175-2.ch004

Das, S. (2021a). *Digital entertainment: The next evolution in service sector*. Springer Nature.https://doi.org/10.1007/978-981-15-9724-4. Print ISBN 978-981-15-9723-7. Online ISBN 978-981-15-9724-4

Das, S. (2021b). *Search engine optimization and marketing: A recipe for success in digital marketing* (1st ed.). CRC Press. https://doi.org/10.1201/9780429298509

Das, S. (2021c) A systematic study of integrated marketing communication and content management system for millennial consumers. In *Innovations in digital branding and content marketing* (pp. 91–112). IGI Global. https://doi.org/10.4018/978-1-7998-4420-4.ch005

Das, S., & Mondal, S. (2016). *Yours digitally: Patanjali*. Available at SSRN 2737299.

Das, S., Mondal, S. R., & Sandhu, K. (2022a). Music logos drive digital brands: An empirical analysis of consumers' perspective. *Journal of Strategic Marketing*, 1–16. https://doi.org/10.1080/0965254X.2022.2098526

Das, S., Mondal, S., Puri, V., & Vrana, V. (2022b). Structural review of relics tourism by text mining and machine learning. *Marketing, 8*(2), 25–34. https://doi.org/10.5281/zenodo.7358349

Duman, Ü. O., & Das, S. (2021). The dynamic rise of digital brands' market mavens in digital entertainment: A complete know how for curious customers. In *Digital entertainment* (pp. 147–172). Palgrave Macmillan. https://doi.org/10.1007/978-981-15-9724-4_8

Duy, N. T., Mondal, S. R., Van, N. T. T., Dzung, P. T., Minh, D. X. H., & Das, S. (2020). A study on the role of web 4.0 and 5.0 in the sustainable tourism ecosystem of Ho Chi Minh City, Vietnam. *Sustainability, 12*(17), 7140. https://doi.org/10.3390/su12177140

Foss, N. J., & Saebi, T. (2016). Fifteen years of research on business model innovation. *Journal of Management, 43*(1), 200–227. https://doi.org/10.1177/0149206316675927

Garipağaoğlu, B. (2021). Digital revolution during pandemic: A case study from Turkey. *International Journal of Educational Research Review*. https://doi.org/10.24331/ijere.911393

Gupta, D. K., Jena, D., Samantaray, A. K., & Das, S. (2019). HRD climate in selected public sector banks in India. *Revista ESPACIOS, 40*(11).

Huang, M. H., & Rust, R. T. (2018). Artificial intelligence in service. *Journal of Service Research, 21*(2), 155–172. https://doi.org/10.1177/1094670517752459

Jain, S., Jain V., & Das, S. (2018). Relationship analysis between emotional intelligence and service quality with special evidences from Indian banking sector. *Revista ESPACIOS, 39*(33).

Joung, J., Kim, K. H., & Kim, K. (2021). Data-driven approach to dual service failure monitoring from negative online reviews: Managerial perspective. *SAGE Open, 11*(1), 215824402098824. https://doi.org/10.1177/2158244020988249

Kaldeli, E., Lazovik, A., & Aiello, M. (2016). Domain-independent planning for services in uncertain and dynamic environments. *Artificial Intelligence, 236*, 30–64. https://doi.org/10.1016/j.artint.2016.03.002

Kim, N., Lee, H., & Kim, W. (2014). Dynamic patterns of industry convergence: Evidence from a large amount of unstructured data. *SSRN Electronic Journal*. https://doi.org/10.2139/ssrn.2519054

Korkeamäki, L., & Kohtamäki, M. (2020). To outcomes and beyond: Discursively managing legitimacy struggles in outcome business models. *Industrial Marketing Management, 91*, 196–208. https://doi.org/10.1016/j.indmarman.2020.08.023

Lekan, A., Aigbavboa, C., Babatunde, O., Olabosipo, F., & Christiana, A. (2020). Disruptive technological innovations in construction field and fourth industrial revolution intervention in the achievement of the sustainable development goal 9. *International Journal of Construction Management*, 1–12.https://doi.org/10.1080/15623599.2020.1819522

Marinova, D., de Ruyter, K., Huang, M. H., Meuter, M. L., & Challagalla, G. (2016). Getting smart. *Journal of Service Research, 20*(1), 29–42. https://doi.org/10.1177/1094670516679273

Minh-Nhat, H. O., Majerova, J., & Das, S. (2022a). Taking steps to help the phoenix rise from ashes: A roadmap for digital enterprises to develop new sustainable goals. In *Sustainable development and innovation of digital enterprises for living with COVID-19* (pp. 35–51). Springer. https://doi.org/10.1007/978-981-19-2173-5_3

Minh-Nhat, H. O., Nguyen, H. L., & Mondal, S. R. (2022b). Digital transformation for new sustainable goals with human element for digital service enterprises: An exploration of factors. In *Sustainable development and innovation of digital enterprises for living with COVID-19* (pp. 85–103). Springer. https://doi.org/10.1007/978-981-19-2173-5_6

Mohanty, P. C., Dash, M., Dash, M., & Das, S. (2019). A study on factors influencing training effectiveness. *Revista Espacios, 40*, 7–15. http://www.revistaespacios.com/a19v40n02/19400207.html

Mondal, S. R. (2020a). *Factors influencing store image loyalty and satisfaction from customer perspective an empirical study in retail sector*. Shiksha o Anusandhan University. http://hdl.handle.net/10603/273688

Mondal, S. R. (2020b). A systematic study for digital innovation in management education: An integrated approach towards problem-based learning in Vietnam. In *Digital innovations for customer engagement, management, and organizational improvement* (pp. 104–120). IGI Global. https://doi.org/10.4018/978-1-7998-5171-4.ch006

Mondal, S. R. (2021) A systematic study of new age consumer engagement and exploration for digital entertainment for over-the-top platforms in various digital media. In *Innovations in digital branding and content marketing* (pp. 113–133). IGI Global. https://doi.org/10.4018/978-1-7998-4420-4.ch006

Mondal, S. R., & Das, S. (2021). Examining diabetic subjects on their correlation with TTH and CAD: A statistical approach on exploratory results. In *Machine learning and the internet of medical things in healthcare* (pp. 153–177). Academic Press. https://doi.org/10.1016/B978-0-12-821229-5.00007-0

Mondal, S. R., Majerova, J., & Das, S. (2022). Sustainable development and innovation of digital enterprises for living with COVID-19. Springer. https://doi.org/10.1007/978-981-19-2173-5

Mondal, S. R., Virgilio, F. D., & Das, S. (2022). *H.R. analytics and digital H.R. practices*. Springer. https://doi.org/10.1007/978-981-16-7099-2

Mondal, S., Das, S., Musunuru, K., & Dash, M. (2017). Study on the factors affecting customer purchase activity in retail stores by confirmatory factor analysis. *Revista ESPACIOS, 38*(61).

Nadanyiova, M., & Das, S. (2020). Millennials as a target segment of socially responsible communication within the business strategy. *Littera Scripta, 13*(1), 119–134. https://doi.org/10.36708/Littera_Scripta2020/1/8

Ng, I. C., & Vargo, S. L. (2018). Service-dominant logic, service ecosystems and institutions: An editorial. *Journal of Service Management, 29*(4), 518–520. https://doi.org/10.1108/josm-07-2018-412

Ngoc-Vinh, N., Tien-Dung, P., & Das, S. (2022). Digitization of Business; Need for Recombination During COVID-19. In *Sustainable development and innovation of digital enterprises for living with COVID-19* (pp. 1–17). Springer. https://doi.org/10.1007/978-981-19-2173-5_1

Nguyen, T. D. M., Mondal, S. R., & Das, S. (2022). Digital entrepreneurial transformation (det) powered by new normal sustainable developmental goals (n-SDGs): Elixir for growth of country's economy. In *Sustainable development and innovation of digital enterprises for living with COVID-19* (pp. 69–84). Springer. https://doi.org/10.1007/978-981-19-2173-5_5

Nishiura, H., & Mimura, N. (2021). Research agenda of climate change during and after the coronavirus disease 2019 (COVID-19) pandemic. *Journal of Clinical Medicine, 10*(4), 770. https://doi.org/10.3390/jcm10040770

Ojo, T. A. (2022). Digital financial inclusion for women in the fourth industrial revolution. *Africa Review*, 1–26.https://doi.org/10.1163/09744061-20220204

Ravi, S., & Mondal, S. R. (2021). Digital entertainment based do it yourself content & advertisement as a factor of driving force for trust & customer patronage. In *Digital entertainment* (pp. 127–146). Palgrave Macmillan. https://doi.org/10.1007/978-981-15-9724-4_7

Rust, R. T., & Huang, M. H. (2014). The service revolution and the transformation of marketing science. *Marketing Science, 33*(2), 206–221. https://doi.org/10.1287/mksc.2013.0836

Sharma, E., & Das, S. (2020). Measuring impact of Indian ports on environment and effectiveness of remedial measures towards environmental pollution. *International Journal of Environment and Waste Management, 25*(3), 356–380. https://doi.org/10.1504/IJEWM.2019.10021787

Sharma, E., & Das, S. (2021). Integrated model for women empowerment in rural India. *Journal of International Development*, 1–18. https://doi.org/10.1002/jid.3539

Sharma, E., Nigam, N., & Das, S. (2020). Measuring gap in expected and perceived quality of ICT enabled customer services: A systematic study of top ten retailers of India. *International Journal of Applied Systemic Studies, 9*(2), 159–184. https://doi.org/10.1504/ijass.2020.113260

Sick, N., Preschitschek, N., Leker, J., & Bröring, S. (2019). A new framework to assess industry convergence in high technology environments. *Technovation, 84–85*, 48–58. https://doi.org/10.1016/j.technovation.2018.08.001

Singh, L. B., Mondal, S. R., & Das, S. (2020). Human resource practices & their observed significance for Indian SMEs. *Revista ESPACIOS, 41*(07). http://www.revistaespacios.com/a20v41n07/20410715.html

Singh, S., & Das, S. (2018). Impact of post-merger and acquisition activities on the financial performance of banks: A study of Indian private sector and public sector banks. *Revista Espacios Magazine, 39*(26), 25.

Siri, R., Mondal, S. R., & Das, S. (2020). Hydropower: A renewable energy resource for sustainability in terms of climate change and environmental protection. In *The handbook of environmental chemistry*. Springer. https://doi.org/10.1007/698_2020_635

Siri, R., & Das, S. (2021). A study on processing of information storage & use of new age consumers in digital wellness sector through story telling & creating interest. In *Digital entertainment* (pp. 45–63). Palgrave Macmillan. https://doi.org/10.1007/978-981-15-9724-4_3

Tien-Dung, P. H. A. M., Majerova, J., & Das, S. (2022a). Phases of possible recovery of digital enterprises in new normal business for living with COVID-19 times: Opportunities for a new era in sustainable development Goals. In *Sustainable development and innovation of digital enterprises for living with COVID-19* (pp. 19–33). Springer. https://doi.org/10.1007/978-981-19-2173-5_2

Tien-Dung, P., Nguyen, X. D., & Das, S. (2022b). Sustainable goal achievement by digital revolution during and after pandemic; how much one wins and losses: A bird's eye view for future planning. In *Sustainable development and innovation of digital enterprises for living with COVID-19* (pp. 133–147). Springer. https://doi.org/10.1007/978-981-19-2173-5_8

Tiwari, M., & Mondal, S. R. (2022). Technological dynamism of workforce management for effective education service delivery during and after Covid-19. In *H.R. analytics and digital H.R. practices* (pp. 199–215). Palgrave Macmillan. https://doi.org/10.1007/978-981-16-7099-2_8

Turber, S., vom Brocke, J., Gassmann, O., & Fleisch, E. (2014). Designing business models in the era of internet of things. *Advancing the Impact of Design Science: Moving from Theory to Practice*, 17–31. https://doi.org/10.1007/978-3-319-06701-8_2

Urbinati, A., Manelli, L., Frattini, F., & Bogers, M. L. A. M. (2021). The digital transformation of the innovation process: Orchestration mechanisms and future research directions. *Innovation*, 1–21. https://doi.org/10.1080/14479338.2021.1963736

Van, N. T. T., Vrana, V., Duy, N. T., Minh, D. X. H., Dzung, P. T., Mondal, S. R., & Das, S. (2020). The role of human-machine interactive devices for post-COVID-19 innovative sustainable tourism in Ho Chi Minh City, Vietnam. *Sustainability, 12*(22), 9523. https://doi.org/10.3390/su12229523

Vargo, S. L., & Lusch, R. F. (2017). Service-dominant logic 2025. *International Journal of Research in Marketing, 34*(1), 46–67. https://doi.org/10.1016/j.ijresmar.2016.11.001

Westerlund, M., Leminen, S., & Rajahonka, M. (2014). Designing business models for the internet of things. *Technology Innovation Management Review, 4*(7), 5–14. https://doi.org/10.22215/timreview/807

Wirtz, B. W., Schilke, O., & Ullrich, S. (2010). Strategic development of business models. *Long Range Planning, 43*(2–3), 272–290. https://doi.org/10.1016/j.lrp.2010.01.005

Xu, L. D., He, W., & Li, S. (2014). Internet of things in industries: A survey. *IEEE Transactions on Industrial Informatics, 10*(4), 2233–2243. https://doi.org/10.1109/tii.2014.2300753

Yegen, C., & Mondal, S. R. (2021). Sharenting: A new paradigm of digital entertainment of new age parenting and social media. In *Digital entertainment* (pp. 213–231). Palgrave Macmillan. https://doi.org/10.1007/978-981-15-9724-4_11

Zott, C., & Amit, R. (2010). Business model design: An activity system perspective. *Long Range Planning, 43*(2–3), 216–226. https://doi.org/10.1016/j.lrp.2009.07.004

CHAPTER 12

Human-Centric Digitalization for New Sustainable Development: An Explorative Study in Human Resource Management

Subhra R. Mondal and Subhankar Das

Introduction

Digital transformation is transforming an organization, industry, or ecosystem. It involves the use of technology to improve processes and enhance efficiency. Digital transformation is about creating value for all its stakeholders, and it uses technology to enable them to adapt to changing conditions (Mondal et al., 2022a). Technology and disruption are not the only factors determining digital transformation's success. It involves people and processes optimizing themselves using information and technology. While digital transformation is commonly used in a business

S. R. Mondal · S. Das (✉)
The Honors Programme, Southern Star Institute of Management, Duy Tan University, Da Nang, Vietnam
e-mail: subhankardas@duytan.edu.vn

S. R. Mondal
e-mail: subhraranimondal@duytan.edu.vn

S. R. Mondal et al. (eds.), *New Normal in Digital Enterprises*,
https://doi.org/10.1007/978-981-19-8618-5_12

context, it can also be applied to address societal challenges such as air pollution and aging populations (Tiwari & Mondal, 2022). In some countries, digital transformation is already being used to improve the quality of life. For instance, Japan has launched an initiative known as Society 5.0, which aims to improve the lives of its citizens (Mondal & Das, 2021). Digital transformation is the process of transforming an organization's activities and techniques to take advantage of the opportunities presented by the rapid emergence and evolution of digital technologies (Sharma & Das, 2021). Developing new competencies involves being more agile, innovative, and customer-centric. This process can help an organization obtain the advantage of new, often untapped, data sources and opportunities. A comprehensive and deep dive into the various digital initiatives and changes across multiple departments is a must for successful digital transformation (Das, 2021a). The scope of digital transformation can be affected by various factors such as the economy's current state, customers' expectations, and changing societal conditions. In practice, optimizing the end-to-end customer experience is a crucial driver of digital transformation. Further, developing new revenue sources and value ecosystems are also vital factors that will help drive digital transformation.

Digital transformation is a journey that involves multiple goals and multiple intermediary goals. While achieving ubiquitous optimization across various processes and divisions is key to success, building suitable bridges between the numerous parties involved is also key to achieving that journey (Siri & Das, 2021). The human element is vital in any digital transformation journey. It is because people value face-to-face interactions and are not interested in digital technology for everything. Yet, in non-digital transactions, digital transformation can also be a part of the solution to empowering those who interact with them (Ravi & Mondal, 2021). A strategy for digital transformation aims to create capabilities that are fully capable of taking advantage of the opportunities presented by new technologies. A successful digital transformation journey should have a straightforward and staged approach that involves various stakeholder groups (Duman & Das, 2021).

Digital Enterprise Transformation for Holistic Development of Business

Digital business transformation is a process that involves the use of technology in various ways, and it has changed the face of business (Mondal,

2020a). Although digital transformation is often used to describe the multiple realities of the industry, it is also used to refer to changes in the way society and the economy operate (Yegen & Mondal, 2021). For instance, digital transformation is often used to describe changes in government regulations and economic conditions (Das, 2021b). Society's shifts can have significant impacts on organizations. No company, industry, or community area is immune from these changes (Sharma et al., 2020). To properly understand the scope of digital transformation, it is essential always to recognize its various facets.

Competitive Advantage on Digitalized Core Competency for Different Enterprises

Digital transformation refers to a wide range of processes and actions that transform an organization (Mondal, 2021). Doing so involves identifying and managing various factors, such as internal and external factors (Das, 2021c). There are huge differences between different regions and industries when implementing digital transformation. For instance, some areas have various regulations and goals (Das, 2020a). This chapter is mainly focused on digital business transformation. It refers to the shifts in the way organizations think about the human element associated with customers, as it is a critical component in the digital transformation process (Duy et al., 2020). The evolution of technologies and various new digital platforms such as the Internet of Things, cloud computing, and big data have enabled companies to transform themselves through digital transformation (Van et al., 2020). However, despite the various advantages of these new platforms, they are only part of the overall process of transforming an organization.

Digital Transformation and Holistic Connection: Highlight the Peripheral Aspect of Enterprises

The last mile of a digital transformation journey often happens at the edges and is key to successful execution. Sometimes, digital transformation is only focused on the customer experience (Siri et al., 2020). This can be a mistake, leaving out other aspects of the project. The end goals of the business and the customers are the driving factors for the digital transformation agenda (Singh et al., 2020). The organization's role is to connect the dots and bridge the silos to achieve these goals. Despite the focus on the edges, the core capabilities are still being realized at a cultural level. The movement toward the edges is influenced by various factors, such as the evolution of computing paradigms and the decentralization of work models (Sharma & Das, 2020). For data management and analysis, researchers are moving towards digital adoption of cloud computing techniques as need for robust analytical output grows in research fraternity (Das, 2020b).

Despite the apparent shift toward the peripheral factors of enterprises, digital transformation is still not limited to an organization with new organizational models. It requires a leader who can oversee the entire process regardless of how it is organized (Mondal, 2020b). In practice, we often see projects already on the way to a more enterprise-wide approach being carried out in an ad hoc or bottom-up fashion (Nadanyiova & Das, 2020). Doing so could put the organization at risk for long-term success. Digital transformation is not an end. It can be a way to meet other strategic goals such as business growth and innovation and a more agile operating model (Behera et al., 2019; Mohanty et al., 2019).

Digital Transformation and Their Scope Covering Different Areas

a. Digital transformation refers to transforming various aspects of a business, including its operations and human resources departments.
b. A successful digital transformation strategy involves optimizing a business process, a combination of various operations and tasks connected to a specific business goal. This process is often achieved using various tools and techniques.
c. A successful digital transformation strategy involves working with a variety of business models. These models help the company transform its core business while exploring other revenue sources and methods.
d. The various ecosystems that a successful digital transformation strategy involves include the relationships between multiple parties involved. These ecosystems are often built upon the foundations of digital transformation.
e. Business asset management is a process that involves identifying and managing non-tangible assets such as customers and information.

This process is vital for achieving a successful digital transformation project.

f. A thriving culture involves establishing a set of core competencies aligned with the organization's customer-centricity goal. This culture can also be created by acquiring critical expertise in various areas such as digital maturity and leadership. The development and operations teams must work together seamlessly to bring new applications to market fast. It is where the culture and processes come into play.
g. The customer, worker, and partner-centricity of a digital transformation project is critically evaluated. This is because the expectations of all stakeholder groups are crucial to the project's success. The rise of cooperative and collaborative models is expected to create new revenue streams and business models. Technology can also be an enabler for addressing the various human aspects of a digital transformation project, such as employee satisfaction and customer experience enhancement.
h. Digital transformation is not just about disrupting or improving technology alone. It also involves working with various parties to create new business models and revenue streams.

Digital Transformation and Related Disruption of Business Sectors

Digital disruption is a concept that refers to the disruption of an industry or ecosystem due to the emergence of new digital businesses that are disrupting the way people do business (Gupta et al., 2019). These new players have successfully mastered the digital skills needed to compete in the fast-changing world. Disruption is about more than just the disruptive tactics of an existing player (Singh & Das, 2018). It also refers to the shifts in power and relationships when people and businesses are disrupted. Digital disruption is a human phenomenon caused by various factors, such as how people use technology and their expectations for their future (Jain et al., 2018). Although multiple factors can cause digital disruption, they can also be triggered by changes in expectations and behavior. Disruption usually occurs at the last mile of a customer's experience (Mondal et al., 2017). For him, it's essential to keep in mind that the disruption effect can have a broad impact and that digital transformation advice should be considered in context (Das & Mondal, 2016). The idea that digital transformation is focused on the edges seems obvious when we consider the increasing expectations of the customers and the people around them.

Digital Disruption and Technological Changes for Enterprises

Although some technologies are not disruptive, some of these have led to more disruption than others. For instance, social media has led to the rise of mobile computing. The evolution of the Internet of Services and its various components has become a crucial part of digital transformation. Big data analytics is also key to the success of many of these technologies.

The Internet of Things refers to the next generation, but its early days are still very much in their infancy. As such, the Internet of Things concept is not meaningful anymore. The Internet of Things will still be the glue that will enable most transformational evolutions. Despite its shortcomings in consumer applications, the Industrial Internet of Things is already driving industrial transformation. The former is due to the emergence of new technologies that are already disrupting the status quo. For now, we are going to remain fully hybrid. We will eventually integrate digital technologies into our lives.

Digital transformation is a process that involves not only managing the existing business but also building for the future. The biggest disruptive technologies of the next generation will be the Internet of Things, AI, Big Data, and cognitive AI. In the meanwhile, the hybrid stage is already here, where the Industrial Internet and Industry 4.0 are both essential components.

The digital disruption happened in different technological aspects, which are discussed below.

a. Technology-induced innovation is the latest trend that can disrupt and transform the way people work and live—some of the technologies that can disrupt artificial intelligence, the Internet of Things, and blockchain.
b. The convergence of various technologies such as AI, Big Data, and the Internet of Things can create new applications.

c. Contrary to popular belief, technology-induced disruption and customer-focused transformation is unrelated to new technologies. Instead, it is the result of various factors such as the demands of customers and the need for more effective and straightforward solutions.
d. When the Internet wasn't there, digital transformation was catching up with the times. As a result, businesses didn't have another option to provide better and more effective customer support.
e. Induced innovation and innovation-induced disruption are two different disruptive technologies that can occur when applied to the human body and mind. For instance, the development of new medicines and technologies to treat diseases can be disruptive.
f. Induced ecosystem-induced change refers to the various changes in the world's ecosystems, which are often triggered by multiple factors such as natural disasters, political upheavals, and regulatory changes. These can create opportunities for businesses to transform their operations and improve customer service.

This concept of the ecosystem brings us back to the importance of thinking holistically about everything that we consume and create. Everything that an organization does is interconnected, with all its activities and models being linked to one another. Imagine how all its business activities are interrelated and how events and changes in the marketplace impact those activities.

Comprehensive Digital Transformation for Enterprises

Interconnectedness and the various accelerations that require deep enterprise-wide change, digital transformation is often seen as a challenging but necessary step to enable organizations to achieve their core business competencies. This transformation is also expected to trigger various changes in how they operate, such as the need for faster response times to deal with the increasing number of customer and partner demands. While digital transformation is an integral part of any business, it is also essential to get a comprehensive view of the whole picture.

Technological Transformation in Digital Language

a. Even if digital marketing is not an integral part of business activities, it can be used in the context of digital transformation.
b. Despite digital customer behavior's role in the business, it is not yet clear how it will play a disruptive role in the future.
c. The concept of paper as an information resource is no longer a requirement.
d. As digital transformation becomes more prevalent, it is only a matter of time before the distinction between physical and digital worlds is blurred.

Digital Transformation Should Not Be Exaggerated for Enterprises

Digital transformation can affect every aspect of an organization. It can create various changes in the way processes and activities are conducted.

a. Contrary to popular belief, digital transformation is not limited to technology companies or even tech startups. It can also be applied to various industries and sectors.
b. It is also a mistake to look at all the tech companies constantly showing off their digital transformation efforts. While some are disrupting the status quo, most of them are doing very challenging and exciting things.
c. The hype around disruptive companies such as Uber and other tech startups is not without danger. Instead, digital transformation is an integral part of any organization's strategy, and it starts with identifying the right goals and challenges.
d. Despite the speed at which digital transformation is happening, many industries are still transitioning and experimenting with different strategies and methods. As for disruptive companies, they can still be disrupted by any potential evolution.

Digital Transformation and Customer-Centric Approach

The digital customer experience is a critical element of any digital transformation. Its combination of various digital platforms and processes can help organizations improve their offerings and reduce costs. Distribution, marketing, and sales are among the earliest areas often considered when it comes to digital transformation.

Digital Optimization and Transformation of Different Sectors with Risk Optimization

a. A modern information management system is focused on delivering a better customer experience. It includes processes and information that feed into them.
b. Human resources, education, and procurement are critical areas that will be examined in this session.
c. To successfully implement the digital transformation, enterprises must connect all the various aspects of a business. Not only do they have to be interconnected, but they also must be people-oriented.

Misconceptions and Realities of Comprehensive Digital Transformation

Digital transformation is becoming more prevalent in many organizations. However, to reap the benefits of this transformative process, organizations should first focus on the real business challenges they face.

Transformation Realities Are Discussed as Follows

a. The relationship between IT and business is critical, as it will play a crucial role in the digital transformation journey.
b. Regardless of the industry, your customers, partners, and competitors will wait for business to catch up.
c. Digital transformation is usually led from the top, and it involves all the stakeholder parties. It's usually led by the CEO, the Chief Digital Officer, or the CIO.

Digital Transformation Economy by the Enterprises in the Social Platform

The rapid emergence and evolution of social platform technology profoundly affect society. With a clear focus on the customer experience and optimizing costs, digital transformation is expected to become the cornerstone of any digital economy. The emergence and evolution of social platform technologies are expected to accelerate the pace of innovation and accelerate the process of transformation. It will lead to a digital economy but make no mistake: digital transformation is still about the customer experience and the human dimension despite all the hype.

Five Stages of Digital Transformation for Enterprises

In this chapter, the authors will talk about the various facets of the digital transformation economy. This concept refers to different digital technologies to transform multiple sectors.

Stage 1: Social and technological platforms for business enterprises: In 2007, the social platform was introduced, including the following four pillars: cloud, big data, mobility, and social media. It was called the Nexus of Forces, and it refers to the various technologies driving social media analytics and cloud technology (Benbrahim et al., 2021). In terms of their adoption, they are transforming the way businesses operate.

Stage 2: Innovativeness for transformation: The social platform was an offshoot of the client–server model era and the mainframe. Several technologies are still predominantly technological but also focus more on business and customer innovation (Degl'Innocenti et al., 2020).

Stage 3: Digital transformation implementation: As companies start implementing initiatives focused on continuous improvement, they become more focused on creating new business models, developing ecosystems, and building new revenue streams, rather than just reacting to existing ones (Fan & Xie, 2021). The various factors that influence a company's decision-making process, including the customer experience, digital marketing, and innovation, are driving the next generation of digital transformation (Krithya Shakthi, 2021).

Stage 4: Expediting innovativeness: The next few years will see the evolution of disruptive business models and the speed at which they can be implemented. This stage of change will propel enterprises to transform themselves and become more competitive (Liao et al., 2020).

Stage 5: Digital transformation propels the economy in post-COVID-19: The digital industrial revolution is about to happen, and it will have a considerable impact on all industries. The innovations that this phenomenon will enable include the Internet of Things, cognitive computing, and the social platform (Patel & Kore, 2020).

Digital Transformation as the Core of Economy and Role of Enterprises

The impact of digital technology has always been significant. In today's world, it is widely believed that the changes brought about by the rapid emergence and evolution of tech have created exponential growth (Paul et al., 2021). This speed is often felt differently in different industries. The various characteristics of a given industry and the various types of customers and stakeholder types influence it. In most cases, not fast organizations can still get left behind (Scekic et al., 2019). An organization can always expect to experience exponential growth due to changes in its environment (Srinivasa Rao & Vazquez, 2020). Doing so helps prepare for the inevitable changes and positions it for success.

Expediting the Innovation for Digital Transformation

While various technologies have accelerated the pace of change, they are just a fraction of the changes that are yet to come. The actual acceleration of digital transformation is when the changes happen so that the expectations of the people who are leading organizations are also surpassed (Wong et al., 2019). It is what analysts refer to as the digital transformation economy.

Proactive Response and Adaption for the Futuristic Transformation of Enterprises

Agility and continuous adaptation are not enough to meet the demands of a rapidly changing environment (Zhang et al., 2020). This concept proposes a strategy that encourages forward-thinking ideas and techniques, which can be applied in various ways (Zou et al., 2020). This strategy must involve various processes and technologies to achieve its goals to be successful.

Digitalization Transformation and Critical Implementation of Resources

Converting document into digital information is a process that is commonly needed in digital transformation. Digitization has also been utilized in other ways, such as recycling paper into digital formats. Digitization is an integral part of digital transformation projects, and it requires various elements to succeed. Change for transformation should be managed by the top leaders holistically by considering every workforce hierarchy. Cosmetic changes can not be the solution in the long term. Changes from the grassroots level to top brass and their mindset for digitalization must be implemented.

Many changes affect various aspects of an organization, and many of these require change management. Knowing the importance of data and analytics to digital transformation is an opportunity to look back at the options created. It is an opportunity to look at what can be done better and identify what should be done differently. It is not new: as web analytics became widespread, many companies started to see the need for digital transformation in their operations. If we fail to consider the concerns and conditions of the individuals we address, this can lead to failure. The first step in any digital transformation journey is having people on board. Without this, an organization cannot realize a truly transformative digital transformation. If people are not considered, the changes may not be as fast as they should.

Urgency and Change in Intent for Digitalization

There are many types of roadmaps for digital transformation projects. They are intended to serve as guiding principles for the various phases of a project, and they are often focused on specific goals and needs. There is never a one-size-fits-all approach for digital transformation projects, as multiple factors and plans can change along the way. Having a prioritized list is also key to ensuring that the goals are aligned with the overall strategy. It's also important to avoid thinking that organizations are already prepared for digital transformation.

Although digital transformation is an integral part of any organization's strategy, it doesn't equal digital transformation. Instead, it's about combining data and processes to achieve business goals. There are still many gaps in digitizing our existing operations and procedures related to paper carriers. There are also huge gaps between the front and back-office processes. It is an example of why digital transformation is not equal to business goals and why many people still feel that there is still a lot of work to be done in society and business.

Digital transformation is about working across silos. In many cases, it involves reworking an organization's structures to enable people to collaborate and remove specific silos. The debate about who is responsible for digital transformation in an organization is archaic. Even if the CEO or other CXOs play a role, the context still matters. It is also vital that CXOs are digital savvy and have the necessary skills and experience to succeed in digital transformation. It is because there are no one-size-fits-all solutions.

Digital Maturity for Enterprises to Implement Transformational Changes

A successful digital transformation strategy should involve a strategic approach. Even though it seems impossible to develop one, it is necessary to understand the various components and processes involved in the project. Getting the correct answers requires a lot of questions. Each business has its unique characteristics, so it is important to start with a strategy that fits its needs.

In digital transformation, it is essential to remember the basics. Complex and challenging questions can be challenging to answer using frameworks and easy answers. Benchmarks and digital maturity frameworks have value. They help organizations develop capabilities that can enable them to implement digital transformation in a strategic and prioritized way. Digital transformation is not simply about a specific project or process. It involves many incremental steps, requiring a team of experts to implement it. The various stages and phases of digital transformation can have multiple goals, and they should fit within the project's overall purpose.

The goal of digital transformation is to prepare for the changes that will occur in the following years due to new technologies and market conditions. A digital transformation strategy is constantly updated, and it considers all the uncertainties and risks involved in the various stages of the project. It ensures that the process can continuously improve and alter its course.

Policies for Digital Enterprises

a. The basic steps of digital transformation are also discussed in this article. However, they also need to be recognized as separate and distinct. Some people see digital transformation as a series of projects related to one another. In other cases, it can be considered a particular project (Das et al., 2022a). Others, like us, see digital transformation as an all-encompassing process that involves the use of various digital technologies. This transformation aims to enable businesses to fully utilize the transformative capabilities of these new platforms and technologies (Das et al., 2022b). This process can be done incrementally and requires a series of steps.
b. Futuristic looks to recover from the pandemic is critical. The rise of digital transformation was widely acknowledged in 2020 due to the COVID-19 pandemic. It presented various opportunities for organizations and societies to improve their digital capabilities; the sheer number of people who started using digital channels for multiple purposes; the inevitable shift from traditional work environments to remote work and hybrid work (Mondal et al., 2022b).
c. COVID-19 impacts the various measures taken by governments and organizations globally, which is very visible in the global economy (Ngoc-Vinh et al., 2022). One of the most critical consequences of COVID-19 is that organizations start to invest in digital experiences and ecosystems. Even if COVID-19 is only focused on climate change, digital transformation is also becoming more significant (Tien-Dung et al., 2022a). With the increasing number of organizations implementing digital transformation programs, it is expected that these will look at all aspects of business and society (Minh-Nhat et al., 2022a).

According to researchers' predictions, by 2023, over 75% of organizations will have a comprehensive digital transformation (DX) implementation plan (Nguyen et al., 2022). It will enable them to transform their businesses and society. Despite the various digital transformation initiatives already being implemented, it is still clear that digital transformation is still accelerating overall. The importance of a holistic approach has become more apparent in the past years due to the various changes brought about by COVID-19. It may seem like a bit of a step too far for many companies. But, for small organizations, this can be a huge opportunity to transform themselves and become more effective in addressing some of the most critical issues of our time (Minh-Nhat et al., 2022b). In 2021, at least 30% of companies will accelerate digital transformation programs to improve their businesses.

DX and Customer-Centric Approach for Enhancing the Experience: The New Standard Marketing

Most digital transformation projects are focused on the user experience and the customer experience. However, both are equally important and require different approaches. The customer experience doesn't belong to one department or another. It should be approached in a transformational manner involving all the stakeholders (Tien-Dung et al., 2022b). While technologies have affected the customer experience, on the one hand, they have also enabled businesses to transform on the other. It requires a comprehensive and holistic approach to enhance the customer experience. The people dimension is probably the most critical area for businesses regarding the customer experience. A customer-centric approach is the most crucial factor of marketing.

The customer experience is at the core of digital marketing transformation. It is driven by the changing expectations of customers and the need to align digital marketing with other divisions such as customer service and IT. It is also why marketing has no choice but to work together with other divisions. The evolution of the marketing function and the role of the CMO are responsible for the success of digital marketing transformation. The goal of digital transformation is to have a comprehensive approach to customer-centric marketing optimization. It is an area that has been neglected due to the silos and separate ways marketing works. In an increasingly complex and connected environment, digital marketing is no longer possible. Instead, it must be done in a truly holistic way, encompassing all the elements necessary to improve the customer experience in a way that's never been done before. This process-oriented approach is often linked with optimizing the customer experience. It can be achieved through automation and cost efficiencies.

Digital Transformation and Data Information About Customer and Market

Even if digital transformation is at the center of information management, the link between this and other elements of digital transformation is not often made enough. Regardless of the type of information you need, it must be used to make the most of it. It is because it provides insight and helps in making decisions. Information chaos is a lot more than just a problem. It can also be an enabler. Information management is a strategic component that supports digital transformation. It plays a vital role in the success of digital initiatives by helping organizations achieve their goals. Information and knowledge management are also critical components of the three-step process: people, technology, and tools.

Intelligent Ways of Managing Information for Digital Transformation

Managing and securing all the necessary information in today's digital environment is not enough. The key is to provide actionable intelligence and actions to drive business results. With the rise of the Internet of Things and artificial intelligence, intelligent information activation has become more prevalent. Therefore, we talk about it. With increasing organizations adopting information-based strategies, intelligent information management has entered the boardroom. The increasing complexity of the data landscape and the need for actionable insights are critical factors that enterprises should consider when implementing intelligent information management. It isn't just about managing information anymore. It also involves connecting various systems and data sets. The increasing complexity of the data landscape and the need to unlock it faster are factors that enterprises should consider when implementing intelligent information management. There are some ways to keep on the information management, such as

a. AI is the process of extracting meaning from vast amounts of data collected by an Internet of Things device.
b. Data capture and management are crucial to achieving faster, better, and more secure transactions. However, to get the most out of the collected data, the process of storing and managing it needs to be done digitally.
c. There has been an increasing focus on quality and accuracy.

Different businesses have unique characteristics. So, it's essential to consider how digital transformation can affect your industry and how it can help you grow. The retail industry is one of the fastest-changing sectors in the world. It is constantly adapting to meet the needs of its customers. Digital transformation is already prevalent in the retail industry. From the front end to the back office, the various processes related to the retail sector are constantly being updated to accommodate the customer's needs. Even though the shop is still critical to many consumers, they have come to expect more innovative experiences when it comes to their shopping experience. Several other technologies have transformed the retail industry, such as the IoT. According to analysts, this tech is already influencing how stores operate.

Expect more innovations that combine various technologies to improve the way people shop for products continuously. But, more than just new products, retailers are also focused on addressing the needs of their customers. The Industrial Internet is already becoming an integral part of the manufacturing industry, experiencing a digital transformation. Combining various technologies, such as data and industrial technology, can improve processes and increase efficiency. Leaders in manufacturing are constantly looking for ways to improve their operations and enhance their customer-centricity. However, many of them are still focused on implementing new strategies and methods of revenue generation. This strategy is not ideal for manufacturers as they may become disruptive due to the lack of a comprehensive digital transformation plan. Although the

Industrial Internet and Industry 4.0 are emerging, their main challenges are still unclear.

The Industrial Internet of Things is an integral part of the industry 4.0 digital transformation. It is proved by the leading role that it plays in manufacturing. The manufacturing industry is leading the way to adopt the Internet of Things. Most of the time, this industry's success is due to the right strategy and execution. But, to truly succeed, this requires a deep understanding of the various challenges and opportunities.

Conclusion

With digital transformation becoming a hyper-connected reality, the need for hybrid, integrated, and inclusive approaches is becoming more apparent. To successfully execute digital transformation, executives need to have a deep understanding of various domains. A CEO needs to know about the different parts of a business process, including reengineering, cybersecurity, and more. Understanding the impact of digital transformation on various areas is a must-have for executives. People and processes are the two most vital requirements which stand on technology for the successful implementation of digital transformation. It is about taking advantage of the power of digital technologies to improve processes and enhance customer experiences. Even beyond digital technology, digital transformation can support and enhance existing processes. It can also create new business models and alter existing ones.

Contrary to popular belief, digital transformation is not a linear process. It involves responding to the changes brought about by digital technologies in our daily lives. The technologies themselves are not the only factors that make digital transformation happen. The human dimension is also an essential part of this process as it informs us about the unforeseen consequences of our actions and behavior. To successfully implement a digital transformation, we need first to establish a culture focused on what people need.

References

Behera, M., Dash, M., Ayasa Kanta Mohanty, A. K., & Das, S. (2019). Entrepreneurial ecosystem: A geographical upliftment. *Revista Espacios, 40*, 26–36. http://www.revistaespacios.com/a19v40n01/19400126.html

Benbrahim, H., Hachimi, H., & Amine, A. (2021). Deep Transfer learning pipelines with apache spark and keras tensorflow combined with logistic regression to detect COVID-19 in Chest CT images. *Walailak Journal of Science and Technology (WJST), 18*(11). https://doi.org/10.48048/wjst.2021.13109

Das, S. (2020a). *Media impact of advertising on consumer buying behaviour a comparative study of different media*. Shiksha o Anusandhan University. http://hdl.handle.net/10603/273691

Das, S. (2020b). Innovations in digital banking service brand equity and millennial consumerism. In *Digital transformation and innovative services for business and learning* (pp. 62–79). IGI Global. https://doi.org/10.4018/978-1-7998-5175-2.ch004

Das, S. (2021a). *Digital entertainment: The next evolution in service sector*. Springer Nature. https://doi.org/10.1007/978-981-15-9724-4. Print ISBN 978-981-15-9723-7. Online ISBN 978-981-15-9724-4

Das, S. (2021b). *Search engine optimization and marketing: A recipe for success in digital marketing* (1st ed.). CRC Press. https://doi.org/10.1201/9780429298509

Das, S. (2021c). A systematic study of integrated marketing communication and content management system for millennial consumers. In *Innovations in digital branding and content marketing* (pp. 91–112). IGI Global. https://doi.org/10.4018/978-1-7998-4420-4.ch005

Das, S., & Mondal, S. (2016). *Yours digitally: Patanjali*. Available at SSRN 2737299.

Das, S., Mondal, S. R., & Sandhu, K. (2022a). Music logos drive digital brands: an empirical analysis of consumers' perspective. *Journal of Strategic Marketing*, 1–16. https://doi.org/10.1080/0965254X.2022.2098526

Das, S., Mondal, S., Puri, V., & Vrana, V. (2022b). Structural review of relics tourism by text mining and machine learning. *Marketing*, *8*(2), 25–34. https://doi.org/10.5281/zenodo.7358349

Degl'Innocenti, M., Frigerio, M., & Zhou, S. (2020). Development banks and the syndicate structure: Evidence from a world sample. *SSRN Electronic Journal*. https://doi.org/10.2139/ssrn.3615477

Duman, Ü. O., & Das, S. (2021). The dynamic rise of digital brands' market mavens in digital entertainment: A complete know how for curious customers. In *Digital entertainment* (pp. 147–172). Palgrave Macmillan. https://doi.org/10.1007/978-981-15-9724-4_8

Duy, N. T., Mondal, S. R., Van, N. T. T., Dzung, P. T., Minh, D. X. H., & Das, S. (2020). A study on the role of Web 4.0 and 5.0 in the sustainable tourism ecosystem of Ho Chi Minh City, Vietnam. *Sustainability*, *12*(17), 7140. https://doi.org/10.3390/su12177140

Fan, L., & Xie, J. (2021). Identify determinants of container ship size investment choice. *Maritime Policy & Managemen*, 1–16. https://doi.org/10.1080/03088839.2021.1971784

Gupta, D. K., Jena, D., Samantaray, A. K., & Das, S. (2019). HRD climate in selected public sector banks in India. *Revista ESPACIOS*, *40*(11).

Jain, S., Jain, V., & Das, S. (2018). Relationship analysis between emotional intelligence and service quality with special evidences from Indian banking sector. *Revista ESPACIOS*, *39*(33).

Krithya Shakthi, K. K. (2021). The Covid-19 pandemic and the global economy: A synopsis. *Global Journal for Research Analysis*, 65–67. https://doi.org/10.36106/gjra/1207748

Liao, X., Wang, B., & Kang, Y. (2020). Novel coronavirus infection during the 2019–2020 epidemic: Preparing intensive care units—The experience in Sichuan Province, China. *Intensive Care Medicine, 46*(2), 357–360. https://doi.org/10.1007/s00134-020-05954-2

Minh-Nhat, H. O., Majerova, J., & Das, S. (2022a). Taking steps to help the phoenix rise from ashes: A roadmap for digital enterprises to develop new sustainable goals. In *Sustainable Development and Innovation of Digital Enterprises for Living with COVID-19* (pp. 35–51). Springer. https://doi.org/10.1007/978-981-19-2173-5_3

Minh-Nhat, H. O., Nguyen, H. L., & Mondal, S. R. (2022b). Digital transformation for new sustainable goals with human element for digital service enterprises: An exploration of factors. In *Sustainable Development and Innovation of Digital Enterprises for Living with COVID-19* (pp. 85–103). Springer. https://doi.org/10.1007/978-981-19-2173-5_6

Mohanty, P. C., Dash, M., Dash, M., & Das, S. (2019). A study on factors influencing training effectiveness. *Revista Espacios, 40*, 7–15. http://www.revistaespacios.com/a19v40n02/19400207.html

Mondal, S. (2021). A systematic study of new age consumer engagement and exploration for digital entertainment for over-the-top platforms in various digital media. In *Innovations in digital branding and content marketing* (pp. 113–133). IGI Global. https://doi.org/10.4018/978-1-7998-4420-4.ch006

Mondal, S. R. (2020a). *Factors influencing store image loyalty and satisfaction from customer perspective an empirical study in retail sector*. Shiksha o Anusandhan University. http://hdl.handle.net/10603/273688

Mondal, S. R. (2020b). A systematic study for digital innovation in management education: An integrated approach towards problem-based learning in

Vietnam. In *Digital innovations for customer engagement, management, and organizational improvement* (pp. 104–120). IGI Global. https://doi.org/10.4018/978-1-7998-5171-4.ch006

Mondal, S. R., & Das, S. (2021). Examining diabetic subjects on their correlation with TTH and CAD: A statistical approach on exploratory results. In *Machine learning and the internet of medical things in healthcare* (pp. 153–177). Academic Press. https://doi.org/10.1016/B978-0-12-821229-5.00007-0

Mondal, S. R., Das, S., Musunuru, K., & Dash, M. (2017). Study on the factors affecting customer purchase activity in retail stores by confirmatory factor analysis. *Revista ESPACIOS, 38*(61).

Mondal, S. R., Virgilio, F. D., & Das, S. (2022a). *HR analytics and digital HR practices*. Springer. https://doi.org/10.1007/978-981-16-7099-2

Mondal, S. R., Majerova, J., & Das, S. (2022b). Sustainable Development and Innovation of Digital Enterprises for Living with COVID-19. Springer. https://doi.org/10.1007/978-981-19-2173-5

Nadanyiova, M., & Das, S. (2020). Millennials as a target segment of socially responsible communication within the business strategy. *Littera Scripta, 13*(1), 119–134. https://doi.org/10.36708/Littera_Scripta2020/1/8

Ngoc-Vinh, N., Tien-Dung, P., & Das, S. (2022). Digitization of Business; Need for Recombination During COVID-19. In *Sustainable Development and Innovation of Digital Enterprises for Living with COVID-19* (pp. 1–17). Springer. https://doi.org/10.1007/978-981-19-2173-5_1

Nguyen, T. D. M., Mondal, S. R., & Das, S. (2022). Digital entrepreneurial transformation (DET) powered by new normal sustainable developmental goals (n-SDGs): Elixir for growth of country's economy. In *Sustainable Development and Innovation of Digital Enterprises for Living with COVID-19* (pp. 69–84). Springer. https://doi.org/10.1007/978-981-19-2173-5_5

Patel, D., & Kore, S. A. (2020). Artificial intelligence: Future impacts, challenges and recommendations on healthcare services. *International Journal of Community Medicine And Public Health, 7*(4), 1596. https://doi.org/10.18203/2394-6040.ijcmph20201480

Paul, D., Sanap, G., Shenoy, S., Kalyane, D., Kalia, K., & Tekade, R. K. (2021). Artificial intelligence in drug discovery and development. *Drug Discovery Today, 26*(1), 80–93. https://doi.org/10.1016/j.drudis.2020.10.010

Ravi, S., & Mondal, S. R. (2021). Digital entertainment based do it yourself content & advertisement as a factor of driving force for trust & customer patronage. In *Digital entertainment* (pp. 127–146). Palgrave Macmillan. https://doi.org/10.1007/978-981-15-9724-4_7

Scekic, O., Nastic, S., & Dustdar, S. (2019). Blockchain-supported smart city platform for social value co-creation and exchange. *IEEE Internet Computing, 23*(1), 19–28. https://doi.org/10.1109/mic.2018.2881518

Sharma, E., & Das, S. (2020). Measuring impact of Indian ports on environment and effectiveness of remedial measures towards environmental pollution. *International Journal of Environment and Waste Management, 25*(3), 356–380. https://doi.org/10.1504/IJEWM.2019.10021787

Sharma, E., & Das, S. (2021). Integrated model for women empowerment in rural India. *Journal of International Development*, 1–18. https://doi.org/10.1002/jid.3539

Sharma, E., Nigam, N., & Das, S. (2020). Measuring gap in expected and perceived quality of ICT enabled customer services: A systematic study of top ten retailers of India. *International Journal of Applied Systemic Studies, 9*(2), 159–184. https://doi.org/10.1504/ijass.2020.113260

Singh, L. B., Mondal, S. R., & Das, S. (2020). Human resource practices & their observed significance for Indian SMEs. *Revista ESPACIOS, 41*(07). http://www.revistaespacios.com/a20v41n07/20410715.html

Singh, S., & Das, S. (2018). Impact of post-merger and acquisition activities on the financial performance of banks: A study of Indian private sector and public sector banks. *Revista Espacios Magazine, 39*(26), 25.

Siri, R., Mondal, S. R., Das, S. (2020). Hydropower: A renewable energy resource for sustainability in terms of climate change and environmental protection. In *The Handbook of Environmental Chemistry*. Springer. https://doi.org/10.1007/698_2020_635

Siri, R., & Das, S. (2021). A study on processing of information storage & use of new age consumers in digital wellness sector through story telling & creating interest. In *Digital entertainment* (pp. 45–63). Palgrave Macmillan. https://doi.org/10.1007/978-981-15-9724-4_3

Srinivasa Rao, A. S. R., & Vazquez, J. A. (2020). Identification of COVID-19 can be quicker through artificial intelligence framework using a mobile phone-based survey when cities and towns are under quarantine. *Infection Control & Hospital Epidemiology, 41*(7), 826–830. https://doi.org/10.1017/ice.2020.61

Tien-Dung, P. H. A. M., Majerova, J., & Das, S. (2022a). Phases of possible recovery of digital enterprises in new normal business for living with COVID-19 Times: Opportunities for a new era in sustainable development goals. In *Sustainable Development and Innovation of Digital Enterprises for Living with COVID-19* (pp. 19–33). Springer. https://doi.org/10.1007/978-981-19-2173-5_2

Tien-Dung, P., Nguyen, X. D., & Das, S. (2022b). Sustainable goal achievement by digital revolution during and after pandemic; How much one wins and losses: A bird's eye view for future planning. In *Sustainable Development and Innovation of Digital Enterprises for Living with COVID-19* (pp. 133–147). Springer. https://doi.org/10.1007/978-981-19-2173-5_8

Tiwari, M., & Mondal, S. R. (2022). Technological dynamism of workforce management for effective education service delivery during and after Covid-19. In *HR analytics and digital HR practices* (pp. 199–215). Palgrave Macmillan. https://doi.org/10.1007/978-981-16-7099-2_8

Van, N. T. T., Vrana, V., Duy, N. T., Minh, D. X. H., Dzung, P. T., Mondal, S. R., & Das, S. (2020). The role of human-machine interactive devices for post-COVID-19 innovative sustainable tourism in Ho Chi Minh City, Vietnam. *Sustainability, 12*(22), 9523. https://doi.org/10.3390/su12229523

Wong, Z. S., Zhou, J., & Zhang, Q. (2019). Artificial Intelligence for infectious disease Big Data Analytics. *Infection, Disease & Health, 24*(1), 44–48. https://doi.org/10.1016/j.idh.2018.10.002

Yegen, C., & Mondal, S. R. (2021). Sharenting: A new paradigm of digital entertainment of new age parenting and social media. In *Digital entertainment* (pp. 213–231). Palgrave Macmillan. https://doi.org/10.1007/978-981-15-9724-4_11

Zhang, G. Z., Deng, Y. J., Xie, Q. Q., Ren, E. H., Ma, Z. J., He, X. G., Gao, Y. C., & Kang, X. W. (2020). Sirtuins and intervertebral disc degeneration: Roles in inflammation, oxidative stress, and mitochondrial function. *Clinica Chimica Acta, 508*, 33–42. https://doi.org/10.1016/j.cca.2020.04.016

Zou, F. W., Tang, Y. F., Liu, C. Y., Ma, J. A., & Hu, C. H. (2020). Concordance study between IBM Watson for Oncology and real clinical practice for cervical cancer patients in China: A retrospective analysis. *Frontiers in Genetics, 11*. https://doi.org/10.3389/fgene.2020.00200

Index

S. R. Mondal et al. (eds.), *New Normal in Digital Enterprises*,
https://doi.org/10.1007/978-981-19-8618-5

The manufacturer's authorised representative in the EU is Springer Nature Customer Service Centre GmbH, Europaplatz 3, 69115 Heidelberg, Germany. If you have any concerns regarding our products, please contact ProductSafety@springernature.com

Printed and bound by CPI Group (UK) Ltd, Croydon, CR0 4YY

30/09/2026

02225243-0003